Transport Research Laboratory

State-of-the-Art Review

Repair of concrete bridges

G. P. Mallett

Thomas Telford

Published by Thomas Telford Services Ltd, Thomas Telford House, 1 Heron Quay, London E14 4JD

First published 1994
Reprinted 1996

Distributors for Thomas Telford books are
USA: American Society of Civil Engineers, Publications Sales Department, 345 East 47th Street, New York, NY 10017–2398
Japan: Maruzen Co. Ltd, Book Department, 3–10 Nihonbashi 2-chome, Chuo-ku, Tokyo 103
Australia: DA Books and Journals, 11 Station Street, Mitcham 2131, Victoria

A catalogue record for this book is available from the British Library

Classification
Availability: Unrestricted
Content: Current best practice
Status: Unrefereed
Users: Practising engineers, academics and students

ISBN 0 7277 2007 4

The opinions expressed in this review are not necessarily those of the Department of Transport.

None of the reported information is to be taken to imply the endorsement of any particular product, method or practice by:

Department of Transport
Scottish Office
Welsh Office
Department of the Environment for Northern Ireland
Transportation Research Board
American Association of State Highway and Transportation Officials
Federal Highway Administration

or any other body, department or organization.

Typeset in Great Britain by Apek Typesetters Ltd., Nailsea, Bristol
Printed and bound in Great Britain by Redwood Books, Trowbridge, Wiltshire.

Cover photograph: M25 Swanley Interchange, Courtesy of Ove Arup & Partners.

Acknowledgements

This review was prepared under contract to the Transport Research Laboratory. The author wishes to thank Dr G. P. Tilly, formerly Head of Structures Division, and Mr A. J. J. Calder, the Project Officer, for their encouragement and advice. The skill of my technical artist, Mr Graham Edwards, in preparing all of the artwork, is much appreciated. Thanks to Mr D. B. Storrar, Bridges Engineering Division of the Department of Transport, for several briefings on recent developments. I gratefully acknowledge the permission of those named in captions to reproduce their illustrations, the assistance of all those who have supplied information in private correspondence and the forbearance of the many authors to whom there are references in the text with extracts, quotations or summaries of their work. The ready cooperation of the following people deserves special mention.

Mr P. Bartle, for guidance on CEB work
Dr N. Buenfeld, Imperial College
Dr J. H. Bungey, University of Liverpool
Mr B. E. Cox, UK/SHRP Coordinator
Mr J. J. Darby, Oxfordshire County Council
Mr E. W. Dore, SCOSS
Dr P. C. Hewlett, Director, British Board of Agrément
Dr G. John, CAPSIS, Manchester
Dr G. C. Mays, Royal Military College of Science
Dr J. F. A. Moore
Dr A. Owens, Strainstall Engineering Services, Ltd
Professor C. L. Page, Aston University
Mr P. Pullar-Strecker
Dr R. J. Reilly, Director, NCHRP, Washington, DC
Dr G. Somerville, BCA
Mr B. S. Wyatt

Finally, thanks are due to my daughter, Jacky, for translating the German paper (2.2.14), to my son, David, for help with Wordperfect and, by no means least, to my wife, Margaret, without whose flexibility on domestic priorities the work would have been neither started nor finished.

Foreword

Concrete bridges form an essential part of the highway infrastructure in the UK, but many have deteriorated significantly over the past 20 years. This is mainly due to corrosion caused by the ingress of deicing salts and alkali-silica reaction. In 1989, the Department of Transport published the results of a survey of 200 concrete bridges, from which it was predicted that a total of about £600 million would need to be spent over the next ten years on maintenance and repair of the country's bridge stock. It is important that any repair option be durable, cost effective, and cause minimum disruption to traffic, as well as complying with the increasingly stringent environmental regulations.

Although concrete repair is a relatively new science, there has been considerable research both in the UK and elsewhere into the effectiveness of different repair methods, and the Transport Research Laboratory (TRL) has been a major contributor to this research. In order to bring together all existing knowledge the TRL commissioned a state-of-the-art review. This contains a wealth of information from a variety of sources including published reports and conference proceedings; it also brings together the accumulated experience of engineers and researchers in the field, and is a unique reference document covering all aspects of the repair of concrete bridges. The various methods are illustrated with numerous case studies and examples of repair strategies.

This review will provide a valuable guide for practising engineers involved in the repair and maintenance of concrete highway bridges, and a useful reference document for researchers and students with an interest in the subject.

John Wootton
Chief Executive, Transport Research Laboratory

About the Author

After education at Hampton Grammar School and graduation from Imperial College in 1945, the Author worked for a short time at RAE, Farnborough, before joining the civil engineering department of the Great Western Railway. Changing to local government in 1955, he was engaged on structural design and was responsible for layout and detailed design of the flyovers at the A3 trunk road junction with Burlington Road and Bushey Road. In 1961 he joined the Bridges Branch of the Ministry of Transport to work on the technical approval of various bridge designs until his promotion in 1966 to the newly formed Design Standards Division. He dealt with bridge loading and standards for concrete highway structures and ancillaries and, in 1970, he was promoted to Superintending Engineer. From 1976 to 1983 he headed the branch dealing with the inspection and maintenance of bridges, which included setting up the bridge database, and contributed to the work of relevant BSI, Research and Liaison Committees. He became Head of Bridges Engineering Division in 1984 and retired in 1986. He was author of *State-of-the-Art Review 2: Fatigue of reinforced concrete*.

Preface

During the 1970s there were reports of massive deterioration of stocks of bridges, predominantly due to salt used on the roads for winter deicing or from sea water. Repair techniques and strategies had to be developed to deal with these special problems of bridges. Over the last decade or so, the growing interest in the durability and maintenance of concrete bridges has produced a wealth of conference papers, research reports and technical and national press reports. Information thus presented has been summarized in this review, including the research at TRL and other centres on deterioration mechanisms, detection of defects, and techniques to repair or protect concrete bridges. The selection of case histories is worldwide, although the majority do come from the UK, and illustrates various problems and solutions. Students will gain a broad view of the subject and references will also be of use to practising academics, engineers and researchers seeking further details.

Contents

Glossary

G1 Bridge management terms

Assessment: evaluation of load carrying capacity, determination of material strength or classification of current condition.

Diagnosis: determining the cause of deterioration.

Improvement: operations to raise the service level above that intended in the original design, e.g. widening, strengthening.

Maintenance: operations to safeguard the integrity of the structure and preserve it against deterioration, e.g. repair, protection.

Protection: operations intended to prevent damage to the structure.

Rehabilitation, refurbishment or renovation: restoration of the service level intended in the original design of the structure.

Repair: the technical aspect of rehabilitation.

Replacement: the ultimate repair of the structure or a part of the structure.

Strengthening: increasing the load carrying capacity of the structure to a level higher than that intended in the original design.

G2 Abbreviations

AAR	alkali-aggregate reaction
ASR	alkali-silica reaction
CMA	calcium magnesium acetate
CSE	saturated copper/copper sulphate electrode
EPS	expanded polystyrene
FB	footbridge
FBECR	fusion-bonded epoxy-coated reinforcement
HAC	high alumina cement
GGBFS	ground granulated blast-furnace slag
GRP	glass fibre reinforced plastic
LWAC	lightweight aggregate concrete
MMA	methylmethacrylate
NDT	non-destructive testing
OPC	ordinary Portland cement
PFA	pulverized fuel ash

pH	logarithmic measure of hydrogen ion activity: $pH < 7$, acidic; $pH = 7$, neutral; $pH > 7$, alkaline
PVC	polyvinyl chloride
SBR	styrene butadiene rubber
SCE	saturated calomel electrode
SHE	standard hydrogen electrode
SSCE	silver/silver chloride electrode

G3 Electrochemical terms

A comprehensive glossary is given as Appendix A to the TRRL Application Guide AG9 (1991), including the following.

Anode: an electrode or part of a metal surface in an electrochemical system where there is a predominance of oxidation reactions. In a corrosion cell, it is the negative electrode at which the corrosion reaction (e.g. $Fe = Fe^{2+} + 2e$) predominates.

Cathode: an electrode or part of a metal surface in an electrochemical system where there is a predominance of reduction reactions. In a corrosion cell, it is the positive electrode at which the reduction reaction (in concrete usually $O_2 + 2H_2O + 4e = 4OH^-$) predominates.

Electrode: a conductor whereby an electrical current passes into or out of an electrolyte.

Half-cell (electrode): one of the electrodes and its immediate environment in an electrochemical cell.

Potential (electrode potential): by convention, an electrode potential is the difference in potential between that electrode and an arbitrary zero, standard hydrogen electrode.

Polarization (polarized potential): departure of an electrode from its equilibrium or steady-state potential; cathodic polarization and anodic polarization cause the potential to become more negative or positive, respectively.

G4 Organizations and bodies

AME	Association of Municipal Engineers
BBA	British Board of Agrément
BCA	British Cement Association (formerly C&CA)
BINDT	British Institute of Non-Destructive Testing
BRE	Building Research Establishment
BRITE	Basic Research in Industrial Technologies for Europe
BSI	British Standards Institution
C&CA	Cement and Concrete Association (now BCA)
CBDG	Concrete Bridge Development Group

CEB	Comité Euro-International Béton
CEN	Comité Européen de Normalisation
CIRIA	Construction Industry Research and Information Association
CRA	Concrete Repair Association
DAfSTb	German Committee on Reinforced Concrete
DGZfP	German Institute of NDT
DoT	Department of Transport
EC	European Commission
FHA	Federal Highway Administration
FERFA	Federation of Epoxy Resin Formulators and Applicators
FIP	Fédération Internationale de la Précontrainte
IABSE	International Association for Bridge and Structural Engineering
ICE	Institution of Civil Engineers
IHT	Institition of Highways and Transportation
IPG	Infrastructure Policy Group of the ICE
IStructE	Institution of Structural Engineers
JSCE	Japan Society of Civil Engineers
LCPC	Laboratoire Central des Ponts et Chaussées
LINK	A linkup of SERC with Government Departments to fund jointly a programme of research on construction maintenance and refurbishment
NCHRP	National Cooperative Highway Research Program
OECD	Organisation for Economic Co-operation and Development
PIARC	Permanent International Association of Road Congresses
RILEM	International Union of Testing and Research Laboratories for Materials and Structures
SCA	Sprayed Concrete Association
SCOSS	Standing Committee on Structural Safety
SCPRC	Society for Cathodic Protection of Reinforced Concrete
SERC	Science and Engineering Research Council
SETRA	Service d'Etudes Techniques des Routes et Autoroutes
SHRP	Strategic Highway Research Program
TRL	Transport Research Laboratory (TRRL before 1992)
TRRL	Transport and Road Research Laboratory (now TRL)

1

Durability and maintenance

1.1 Introduction

1.1.1 Scope

Over the last decade, there has been considerable interest in the durability and maintenance of concrete bridges and a wealth of knowledge has been promulgated at various meetings or in papers, books or press reports. This State-of-the-Art Review is concerned solely with the repair of deterioration or damage. The aim is to outline the present position and the latest developments in the field of repair and give references where details may be found; no attempt is made to reproduce full details. It does not deal with design, assessment of load carrying capacity, strengthening or other improvements, although these are often associated with repairs, or may even spark the need for repairs. Codes which have been issued by the British Standards Institution (1984) for design and by the Department of Transport (1984) for the assessment and strengthening of highway structures may be relevant to some aspects of repair.

1.1.2 Deterioration of bridges

In the 1970s there was a growing awareness in many countries of the need to safeguard their stocks of bridges as they are key elements in their highway systems. In France, for instance, the Ministère de l'Equipment (1975) issued an illustrated guide to classify the type and severity of the various defects in concrete bridges.

In the United States the position was particularly acute and the Federal Highway Administraion (1970) had issued a training manual for bridge inspectors. *Engineering News Record* (1977) reported that more than one in six bridges were deficient and that the 40 000 deficient bridges on the federal-aid system would take more than 200 years to replace at the rate that funds were being allocated.

Kulash (1990) described the goals and recent developments in concrete and structures in the Strategic Highway Research Program (SHRP). The two major goals were: improving durability of concrete in highway structures; rehabilitation and protection of concrete bridges.

The work included developments in:

- ☐ prevention and repair of damage due to alkali-silica reaction, freeze-thaw, corrosion,
- ☐ quality control and quality assurance,
- ☐ non-destructive testing,
- ☐ use of admixtures,
- ☐ condition assessment of components,
- ☐ electrochemical repair and protection techniques,
- ☐ chemical corrosion inhibitors,
- ☐ conventional physical repair methods.

Ahlskog (1990) reported that 42% of the 578 000 public highway bridges in the United States were classified as deficient by the Federal Highway Administration. He discussed the past, current and future bridge management systems.

The significant technical and economic problems faced by road authorities in many countries were addressed by the OECD Road Research Programme as a result of which OECD published a series of reports:

- ☐ Waterproofing of concrete bridge decks (1972)
- ☐ Bridge inspection (1976)
- ☐ Evaluation of load carrying capacity of bridges (1979)
- ☐ Bridge maintenance (1981)
- ☐ Bridge rehabilitation and strengthening (1983)
- ☐ Durability of concrete road bridges (1989)
- ☐ Bridge management (1992)

The latest report reviewed experience in the development of Bridge Management Systems (BMS) in OECD member countries. It dealt with optimization of inspection and of maintenance, prediction of service life, bridge data banks and BMS. Blakelock (1993) described the development of a computerized BMS and its use by a number of authorities in the UK.

The penultimate report outlined durability phenomena, economic issues and the current position in the OECD countries. It analysed causes and mechanisms of deterioration and reviewed diagnostic and repair techniques not covered in the earlier reports. General principles for durability technology include structural reliability of deteriorating bridges, predicting rates of deterioration and some economic aspects of bridge management. Siemes and Vrouwenvelder (1985) had proposed the application of reliability to durability design. They found that this is feasible and reported a recent decision in The Netherlands to prepare a probability based code of practice for durability, not for mandatory general application but to be available for special cases. Miyagawa (1991) postulated a unified approach to durability design and repair based on the mechanism of deterioration.

In the United Kingdom, waterproofing of bridge decks has been mandatory since 1965 and standards are revised from time to time in the light of experience. Price (1989) investigated the performance of a wide range of available waterproofing systems and found many to be susceptible to damage under prevailing conditions. New standards for bridge inspection were prescribed by the Department of Transport (1977) in a technical memorandum, information from which has, since 1988, been integrated in a computerized data bank. A Bridge Inspection Panel (1983) prepared a guide to assist bridge inspectors. *New Civil Engineer* (1984) reported Vassie (1984) discussing the growing problem of reinforcement corrosion and some methods for detecting it. The Department of Transport (1986 and 1990) issued advice on investigation, repair and protection of concrete highway structures and standards for repair materials. The TRRL published a number of reports which are referenced in the relevant chapters. Wallbank (1989) reported on the condition of 200 randomly selected concrete highway bridges; 61 were found to be in poor condition. The report included estimates of the annual costs of repair and protection of concrete bridges within the 15-year programme announced by the Minister for Roads and Traffic in 1987. If those estimates are expressed as percentages of replacement value of the bridge stock, a buildup was proposed from about 0.6% to more than 3% in the first few years to catch up the repair backlog, after which the annual expenditure could be reduced to its former level. (See also 3.1.1.) Holland and Dawe (1990) described the Department of Transport's 15-year strategy for rehabilitating its highway bridges and outlined the various studies which have been undertaken to support this remedial work. Organization and management of the programme was described, including some of the problems facing a bridge authority in maintaining its bridge stock with limited funds. In fact the DoT (1993) annual expenditure on rehabilitation of concrete bridges did increase to about 1.8% by the fourth year but was constrained by two factors:

- time taken by agents to gear up to a higher rate of work;
- budgetary controls became dominant later.

1.1.3 Durability and maintenance

Concrete bridges are required to maintain their serviceability over long periods of time, hopefully 100 years or more.

Durability aspects of serviceability are represented in the current codes by deterministic rules for concrete cover and crack widths. When fuller information has been gained about rates of deterioration it is conceivable that durability will become a limit state in its own right. If deterioration is neglected it will get worse; it is invariably more economical to repair sooner rather than later and before the need for replacement.

Before deciding on a repair, it is necessary to diagnose the cause of the damage and the degradation mechanism. Only then can corrective maintenance costs be estimated realistically. The process leading to an effective and durable repair is:

- detection
- investigation
- diagnosis

- evaluation of options
- repair.

The following formula, given by Bijen (1989), can be used to evaluate different repair options:

$$E(C_{\text{cap}}) = S + \sum_{j=1}^{n} \frac{V_j}{(1+r)^j} + \sum_{j=1}^{n} \frac{P(F_j)D_j}{(1+r)^j}$$

in which E (C_{cap}) is the expected value of capitalized costs C_{cap}, S represents investment costs, V_j the maintenance and management costs for year j, $P(F_j)$ the probability of failure in year j, D_j the costs for corrective maintenance in year j, r the real interest (interest adjusted for inflation), n the number of years to function without defects.

Steel embedded in concrete is normally passivated, that is coated with oxide, by the strongly alkaline pore water. Deterioration of concrete bridges is frequently due to the corrosion of steel reinforcement where the passivation has been breached. The mechanisms of protection and corrosion were documented by the Building Research Establishment (1982), CEB (1982), Page and Treadaway (1982), Wilkins and Lawrence (1983).

Vassie (1984) distinguished general corrosion and localized corrosion which occur in concrete bridges. The corrosion of the steel is electrochemical, forming anodes and cathodes; either a homogeneous system producing general corrosion or local concentrations forming anodic pits surrounded by the passivated areas which act as cathodes. Bijen (1989) p 30, quoted the Hausman criterion of 0.6 for the chloride/hydroxyl ratio of pore water at the steel surface.

Thomas and Matthews (1991) found PFA concrete to be superior to ordinary Portland cement concrete. (See 1.5.1 and 7.1.1.)

1.2 Corrosion mechanisms

1.2.1 Carbonation

The loss of alkalinity of concrete caused by the entry of carbon dioxide was described by Roberts (1981), Department of Industry (1981), Bijen (1989) pp 14, 15, and Kay (1992) pp 39–41. Martin, Rausen and Schiessl (1975) and Klopfer (1981) gave the following equation for the rate of penetration of CO_2 :

$$x = (2Dt)^{0.5}$$

where x is depth of penetration (mm), D is diffusion coefficient (mm^2/year), t is time (years).

When the carbonation front reaches the reinforcement its passivation is lost and corrosion can start if moisture and oxygen are present. Usually the corrosion is general with expansive rust products causing spalling. The rate of carbonation is highest at about 60% relative humidity and very low at either 25% or 100% relative humidity. It is likely to be associated with old bridges or inadequate depth or quality of concrete cover. Wallbank (1989), in a random sample of 200 highway bridges, found that 90% of carbonation depths were 5 mm or less. Wood (1990) suggested that the worst conditions are wetting and drying with relative humidity between 60 and 90%. Diffusion coefficients are notoriously inconsistent but can be calculated with some confidence where carbonation depths after a number of years are available. (See 3.3.5.)

1.2.2 Chloride

The common source of chloride, if not from the sea, is from deicing salt, either leaking through faulty waterproofing or deck joints or being splashed up by traffic. The presence of chloride ions can cause depassivation of steel even if the associated pore solution is highly alkaline. Pocock and Gibbs (1986) quoted Browne (1982) who plotted chloride distribution curves from samples taken at incremental distances from the concrete surface and found that Fick's diffusion law applied:

$$dC/dt = D_c d^2C/dx^2$$

where C is chloride ion concentration at depth x after time t and D_c is the chloride diffusion coefficient.

Holden, Page and Short (1983), p 144, studied the reaction of chlorides in the mix with different types of cement.

Vassie (1984) dealt specifically with reinforcement corrosion and the durability of concrete bridges exposed to deicing salt contamination.

The complex destructive effect of chloride combined with frost-thaw was described by Bijen (1989), pp 26–28. He also gave polarization curves, Pourbaix diagrams and the diagram of pit corrosion on which Figure 1.1 is based. This localized pitting corrosion, with non-expansive corrosion products, can quite possibly develop without visible signs on the surface of the concrete.

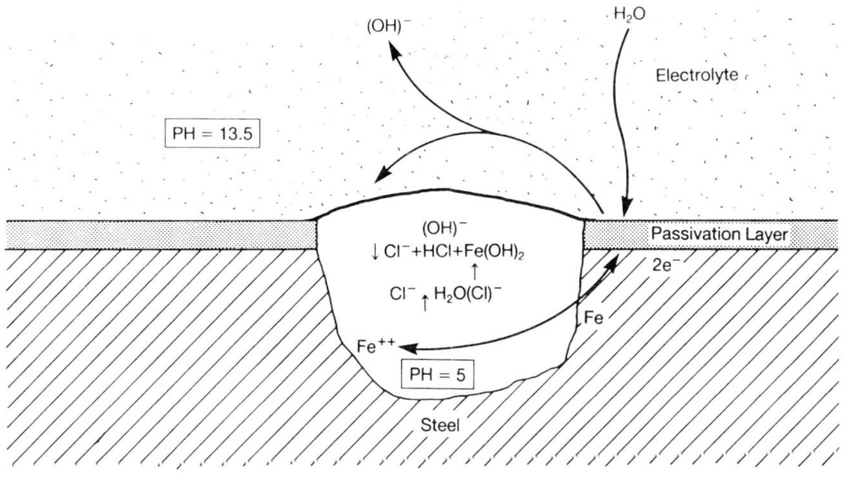

Fig. 1.1. Aggressive environment in chloride corrosion pit (Bijen, 1989)

1.2.3 Sulphate

Sulphate contamination from groundwater or industrial pollution of air or water may have a depassivating effect similar to that of chlorides. Bijen (1989), pp 16, 17, discussed expansive salts formed by sulphates. Holden, Page and Short (1983) found that the combined presence of sulphates and chlorides in the mix resulted in an increased risk of corrosion. Experience of this in road bridges built between 1971 and 1974 in Cambridgeshire was reported in *New Civil Engineer* (1986). Leakage and spray of salty water had caused deterioration of concrete and some general corrosion of steel reinforcement. Concrete surrounding the steel contained less than 0.2% of chloride by weight of cement but concentrations of sulphates in the cement paste had increased from about 2% originally to 6–8% by weight of cement. Although the rock salt used for deicing should contain less than 2% of calcium sulphate to comply with BS3247, concentrations can build up in concrete subjected to alternate wetting and drying. Expansive reactions within the pores and cracks of the concrete result.

1.3 Other damage to concrete

1.3.1 Frost-thaw

Bijen (1989), pp 26–29, described the mechanism of frost damage as water freezes in the pores of concrete and the exacerbating effects of deicing salt. Bennison (1987) related frost damage to porosity or cracking of the concrete and pressures built up under restraint may depress the freezing point down to -20°C.

1.3.2 Alkali-silica reaction

Alkali-silica reaction may result from the combination of:

- reactive aggregate,
- high alkali cement, and
- moisture.

An expansive gel formed in the matrix of the concrete produces characteristic map cracking. ASR had been known for some time in Denmark, interestingly described by Idorn (1967), also in the USA, Canada and South Africa, when it appeared in 1976 in Britain. Cases reported were mainly in Devon, Cornwall and Warwickshire but it was found that precautions were necessary to avoid its occurrence in other parts of the country. The Department of Transport, monitoring its 6000 concrete bridges in 1987, found 165 confirmed cases of ASR and 303 suspect or potential cases.

Guidance on minimizing the risk of alkali-silica reaction prepared in 1983 was subsequently modified and extended in the light of new data and comments and the definitive report was published by the Concrete Society (1987). The Institution of Structural Engineers (1990) issued guidance on the structural effects of ASR.

Reinforcing steel is stressed in tension as it restrains the expansive action, produces compression in the concrete and modifies the cracking. The risk of ASR can be virtually eliminated from new designs. Where it has started, it cannot be arrested completely but it can be slowed down and its damage can usually be repaired rather than resorting to demolition. Vassie (1993) reported that ASR cracking increases the absorption of chloride and therefore tends to exacerbate corrosion of reinforcement although the effect of carbonation in cracks is not significant. The British Concrete Association (1992) has published a definitive bibliography. Clark (1993) discussed the assessment of bridges affected by alkali-silica reaction.

1.3.3 High alumina cement

High alumina cement concrete was used quite widely in the 1950s and 1960s in buildings and to a lesser extent for bridges. Its advantages are high early strength and resistance to sulphate attack but its disadvantage is its long-term conversion to a more stable mineral form with a considerable loss in strength from that of its early form. In 1974, after the collapse of a prestressed HAC concrete roof beam in a school building, warnings and advice were issued to Local Authorities to identify HAC concrete in their buildings and take the necessary action. Similarly the Department of the Environment issued advice (1974) to Highway Authorities to identify and check those of their bridges which had HAC concrete in their construction. Advice included suggestions for testing prepared by Stevens (1974). This advice could be useful in dealing with any HAC concrete which may still be encountered. Loe, Irwen and Grainger (1975) carried out some loading tests on floors with HAC joists to assist the Building Research Establishment in their comprehensive investigation of the problem. Neville and Wainwright (1975) documented the properties of converted HAC. The Institution of Structural Engineers' Working Party Report (1976) gave further useful information. *New Civil Engineer* (1993) reported that BRE had found a number of buildings in which carbonation had reached the steel reinforcement. Corrosion had started in about one-fifth of the structures examined but had not yet led to any cracking or spalling. Now that most HAC concrete is more than 20 years old, BRE is embarking on a new study of its rate of deterioration.

1.3.4 Impact

Impact damage by a vehicle, load or ship travelling under a bridge may range from minor to major, such as:

- local damage to cover concrete of superstructure,
- local damage to cover concrete of substructure,
- shifting superstructure on bearings,
- local damage to superstructure and to bearings,
- serious damage to fascia beam,
- serious damage to bridge support,
- destruction of superstructure.

The first action is to make the road safe by the temporary closure of lanes as necessary and the provision of signs, temporary supports, fenders, handrails, etc. until the permanent repair can be carried out. Each case must be decided separately. Some examples of impact damage and repairs are given in 2.9.

1.3.5 Fire

The Concrete Society has published Technical Report No. 33 (1990) on the assessment and repair of fire-damaged concrete structures. Experience has shown that reinforced concrete structures can nearly always be reinstated after a fire and only in case of severe damage need members be replaced.

The key to the assessment of damage is in the effects of high temperatures on the structural materials. At temperatures up to 300°C, the residual strength of concrete is not severely reduced but above 500°C compressive strength may be reduced to a small fraction of its former value. Pink discolouration is an important sign of the onset of significant loss of strength. Unfortunately the absence of pink discolouration does not necessarily mean that the concrete has not been weakened. As concrete is a poor conductor of heat, quite high temperatures at the surface may result in spalling but no damage only a few millimetres deeper. Malhotra (1984) gave a comprehensive study of spalling in fires. The Concrete Society's Technical Report No. 33 (1990) gives a series of temperature distributions for various concrete cross-sections. It suggests conservative values of the reduced residual yield strength after exposure to temperatures of 450°C to 700°C for cold worked steel and 600°C to 700°C for hot rolled steel which are sufficient for most purposes; it can be assumed that there is no loss at lower temperatures.

The effects of high temperatures on prestressing steels are more critical. The ultimate tensile strength, the residual stress and the elastic and relaxation properties are likely to be adversely affected at temperatures above 200°C.

1.4 Post-tensioning tendons

1.4.1 The durability problem

In the United Kingdom the first post-tensioned bridge was constructed in 1947; post-tensioning was adopted subsequently for more than 600 DoT concrete bridges and possibly 3000 in all in the UK. A few of these had the tendons external to the structural section but generally they were placed in ducts within the section, stressed, anchored and then grouted. Despite the intentions of specifiers to fill the ducts with cement grout completely, this was not always achieved. In one investigation, Woodward (1981) reported voids in ten out of twelve bridges and more than half of the ducts that he examined. It was found that even traces of grout in a voided duct generally provided sufficient alkalinity to prevent corrosion of the tendon in the absence of chlorides. However, sites vulnerable to the ingress of chlorides raise serious concern about the corrosion of the tendons and so require careful assessment followed by monitoring or other appropriate action.

Woodward and Loe (1981) described the hole drilling and vacuum method for monitoring voided ducts. They suggested the development of monitoring techniques as no suitable non-destructive testing techniques were available to detect corrosion or fracture of post-tensioning tendons. Barton and Kusenberger (1981) at the same conference reported magnetic field equipment, with signal enhancement features, to be a promising method. Several post-tensioned bridges have had to be demolished due to serious pitting corrosion of the tendons. Case histories are summarized in 2.2.14, 2.5.7, 14, 15 and 21.

In view of the concerns about the durability of post-tensioned bridges, the DoT decided in 1992:

- ☐ to commission no more grouted post-tensioned bridges, pending the preparation and issue of new standards for design and construction;
- ☐ to intensify the monitoring of its post-tensioned bridges and develop information and training on the subject.

Following a detailed study of the literature relating to structures in service, the Concrete Bridge Development Group (CBDG) is cooperating with the Concrete Society to update the documentation for grouted post-tensioned bridges. With active input from DoT, TRL and the County Surveyors Society (CSS), Task Groups were formed to report on the following:

- ☐ specification for grout and grouting, substantially based on draft CEN standards and supported by a certification scheme for site operatives and site operations,
- ☐ guidance on design and detailing of bridges for durability,
- ☐ evaluation of vacuum grouting,
- ☐ development of reliable test methods to detect any voids soon after grouting is complete.

The close liaison in this activity includes the British Standards Institution (BSI). Clark (1992) presented a review of the literature. The DoT (1993) issued relevant guidance for Special Inspections.

1.4.2 Segmental construction

Many post-tensioned bridges are constructed with precast concrete segments. In the early days it was the practice to dry pack mortar in the thin joints between units prior to stressing. This has proved to be a point of weakness for the entry of salt-laden water, which was in fact the cause of the collapse of the Ynys-y-Gwas bridge in 1953, reported in *New Civil Engineer* (1985), more fully by Woodward and Williams (1988) and further reports by Woodward (1989 and 1991). The later practice of forming joints 100 mm wide with high quality concrete was intended to be more durable, although this remains to be proved.

1.5 Measures to improve durability

1.5.1 Better concrete cover to reinforcement

Brown (1986) reported a study of a selected sample of 92 concrete highway bridges in England and Scotland. The age and area distribution of the sample is given in Figure 1.2.

Bridges constructed more than 50 years ago are termed old; those less than 25 years old are called modern. The results showed that modern concretes have higher cement content and lower water/cement ratio than the old ones. Modern concretes carbonate less rapidly than the old and none of the modern bridges had such low cover as some of the old ones, where carbonation had reached the steel and corrosion had started. Extrapolation of the carbonation depths suggested that they would not exceed 28 mm in the modern bridges after 120 years. Every case of spalling in modern bridges was associated with high chloride concentration.

Current standards require greater depths of cover and better compaction than those generally achieved in older bridges. Research has shown that the use of 65% ground granulated blast-furnace slag (GGBFS) with 35% Portland cement results in low porosity concrete. Also 30% or more of pulverized fuel ash (PFA) as cement replacement gives a denser concrete which is more resistant to the penetration of chloride ions as reported, for instance, by Thomas (1991) and Thomas and Matthews (1991).

Fig. 1.2. Distribution of bridges by date of construction (Brown, 1986, TRRL CR 43)

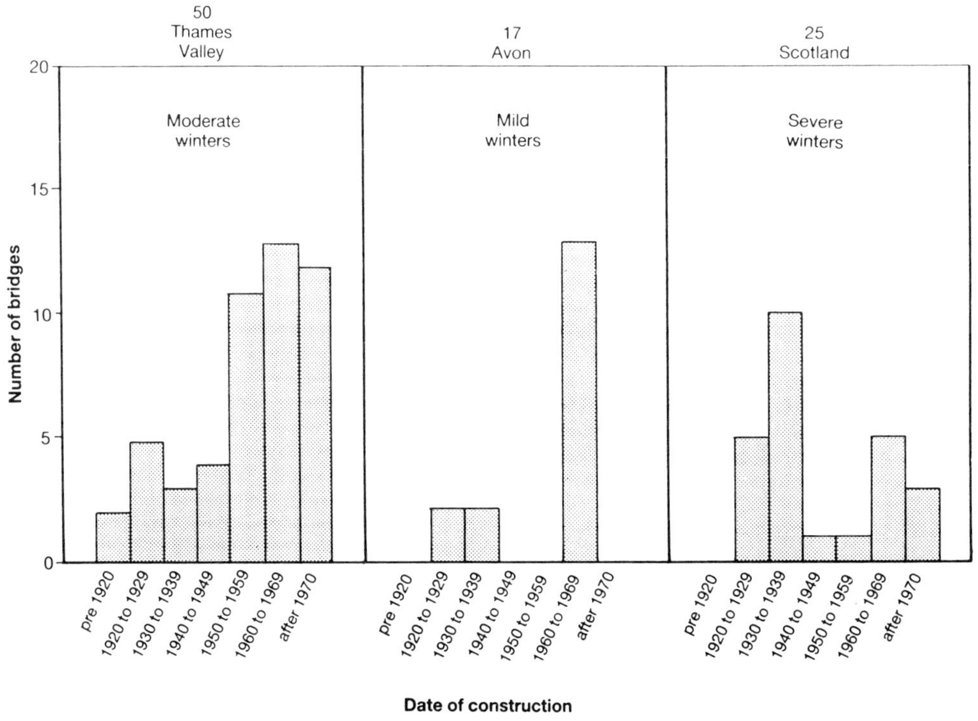

Cabrera, Cusens and Lynsdale (1989) reported an extensive laboratory study from which they obtained statistical models linking oxygen permeability to pore structure, compressive strength, porosity, water/cement ratio, air content and age. From this work, they proposed a value for oxygen permeability at one day as a concrete quality control criterion (7.2.1).

1.5.2 Bridge deck waterproofing

The absence of waterproofing on many American bridge decks is a major factor leading to the scale of their deterioration. Carriageway resurfacing is an opportunity to provide waterproofing on an unprotected bridge or to repair leaking waterproofing and joints. (See also 5.3.2.)

1.5.3 More effective use of deicers

Economy in the use of salt and the use of non-aggressive alternatives to salt are considered in 5.1.3.

1.6 Trends and current research

1.6.1 Equipment

Bungey (1992) works at the University of Liverpool on testing concrete properties in situ. Recent developments by the testing equipment makers are largely concerned with automation, data logging and processing and monitoring systems, such as:

- ☐ half-cell potential multi-cell data logger,
- ☐ Wenner probe resistivity data logger,
- ☐ pulse echo techniques,
- ☐ sub-surface radar detection of bars or other features.

1.6.2 Deterioration mechanisms

Buenfeld (1992) heads a Research Group at Imperial College, London, measuring and modelling molecular transport in concrete. Specific projects are:

- ☐ admixtures to reduce chloride penetration in concrete,
- ☐ wick action in concrete,
- ☐ chloride penetration resistance of surface treatments.

Page (1992) at Aston has current projects on:

- ☐ pore structure and diffusional properties,
- ☐ interface microstructure,
- ☐ degradation and corrosion mechanisms.

Thompson (1992) works at TRL on durability and repair, such as:

- ☐ alkali-silica reaction, structural implications,

- ☐ concrete repair criteria and techniques,
- ☐ durability of new construction.

John (1992) works at CAPSIS, Manchester, on:

- ☐ rates of corrosion,
- ☐ effectiveness of repairs on concrete structures in service.

1.6.3 Repair techniques

Mays (1992) heads the Civil Engineering Group at the Royal Military College of Science, Shrivenham, working on the performance of concrete structures, including:

- ☐ structural effects of patch repairs,
- ☐ polymer mortars,
- ☐ structural adhesives,
- ☐ external plate reinforcement,
- ☐ strengthening reinforced concrete beams in shear,
- ☐ property mismatch in patch repairs.

1.6.4 Lightweight aggregate concrete

Mays and Barnes (1991) investigated lightweight aggregate concrete structures in service and found them to be no less durable than normal weight aggregate structures.

1.6.5 Protection

Thompson (1992) works at TRL on protection, including:

- ☐ cathodic protection,
- ☐ surface treatments for concrete,
- ☐ bridge deck waterproofing.

Page (1992) at the University of Aston and John (1992) at CAPSIS are working on cathodic protection. Page (1992) also has projects on:

- ☐ electrolytic desalination of concrete,
- ☐ use of coatings to control corrosion rates,
- ☐ alkali-silica reaction,
- ☐ deicing agents.

1.7 Information

1.7.1 Books

Relevant information has been published by the following:

Perkins (1986)
Cope (1987)

Pullar-Strecker (1987)
Kay (1992)
Mailvaganam (1992)
Mays (1992)

1.7.2 Some conferences, seminars and symposia

13 – 17 April 1981. *International Conference on Inspection, Maintenance and Repair of Road and Railway Bridges.* Bruxelles-Paris.

6 November 1981. *The Failure and Repair of Corroded Concrete Structures.* One-day Conference at the Europa Hotel, London. Organized by Oyez IBC. Scientific and Technical Studies.

13 – 15 June 1983. *Corrosion of Reinforcement in Concrete Construction.* Papers presented at the meeting at the Kensington Close Hotel, London. Proceedings edited by Alan P. Crane. Society of Chemical Industry. Published by Ellis Horwood.

6 November 1984. *Repair of Concrete Damaged by Reinforcement Corrosion.* One-day conference arranged by the Concrete Society in conjunction with the Association of Gunite Contractors and the Federation of Resin Formulators and Applicators. Tara Hotel, Kensington, London.

25 September 1985. *Concrete Bridges – Investigation, Maintenance and Repair.* Concrete Society. Tara Hotel, Kensington, London.

8 April 1986. *National Workshop on Bridge Maintenance Initiatives.* Institution of Highways and Transportation. Leamington Spa.

16 March 1988. *First Annual Conference on Construction Repair.* London Tara Hotel, Kensington. Palladian Publications Ltd.

23 February 1989. *Concrete bridges—management, maintenance and renovation.* London. Concrete Society (1989a). CS.102.

27–29 June 1989. *International Conference on Structural Faults and Repair,* Kensington Town Hall, London. Proceedings edited by M. C. Forde. Engineering Technics Press.

October 1989. *International Conference on Deterioration and Repair of Reinforced Concrete in the Arabian Gulf.* Bahrain Society of Engineers.

1989. *Symposium on Durability of Structures.* Lisbon. International Association of Bridge and Structural Engineering.

20 March 1990. National Workshop: *New Concepts for the Management of Highway Structures.* Institution of Highways and Transportation. Leamington Spa.

28 – 30 March 1990. *First International Conference on Bridge Management.* University of Surrey, Guildford.

30 April – 2 May 1990. *NATO Advanced Research Workshop on Bridge Evaluation, Repair and Rehabilitation.* Baltimore, Maryland, USA. Proceedings edited by A. S. Nowak. Nato ASI Series E: Applied Sciences Volume 187. ISBN 0 7923 0999 5.

21–24 May 1990. *Third International Conference on Corrosion of Reinforcement in Concrete Construction.* Wishaw, Warwickshire, Society of Chemical Industry. Published by Elsevier Applied Science.

11–13 September 1990. *International Conference on Protection of Concrete.* University of Dundee. Proceedings edited by R. Dhir and J. W. Green.

7 – 9 November 1990. *Fifth International Conference on Durability of Building Materials and Components.* Brighton. E. and F. N. Spon.

12 November 1992. *Durability of Concrete Structures Symposium.* The Institution of Highways and Transportation, South Western Branch. County Hotel, Taunton. Unpublished.

14 –16 April 1993. *International Conference: NDT in Civil Engineering.* British Institute of Non-Destructive Testing. University of Liverpool.

18 – 21 April 1993. *Second International Conference on Bridge Management.* University of Surrey, Guildford.

1993. *Fifth International Conference on Structural Faults and Repair — Extending the Life of Bridges — Civil and Building Structures.* Edinburgh. Proceedings edited by M. C. Forde.

17 February 1994. *Rehabilitation of Concrete Structures Symposium.* The Institution of Highways and Transportation, South Western Branch. County Hotel, Taunton. Unpublished.

1.7.3 British Board of Agrément (BBA)

The BBA assesses mainly innovative proprietary materials and systems which are not covered by British Standards or which deviate from standards covering the same application. An assessment of performance is carried out against a method statement compiled by the BBA with a view to providing an assurance that the material or system will, subject to any conditions stated in the Agrément Certificate, be fit for its stated purpose. The BBA collaborates with other national members of the European Union of Agrément to develop Product Guidelines on Methods of Assessment and Testing. Agrément Certificates have been issued for bridge deck waterproofing systems and for several proprietary concrete repair systems.

1.7.4 British Cement Association (BCA)

The BCA (known, prior to December 1987, as the Cement and Concrete Association) is an independent non-profit-making body financed entirely by the British manufacturers of Portland cement. It provides a service of advice and information to all users of cement and concrete in the UK. The Association's experience has been accumulated from the national and international exchange of technical information and, since 1947, its own research and development activities. The BCA offers a range of courses and shorter events and a Concrete Information Service through Concrete Current Awareness papers published monthly. The BCA disseminates information through its publications, technical reports and papers, meetings and Library Information Service.

1.7.5 Concrete Bridge Development Group (CBDG)

The CBDG was set up in 1992 to promote the greater and better use of concrete in bridges. Membership is open to bodies which subscribe to the objectives set out in the Memorandum and Articles of Association, in three categories:

- Group Membership for industry
- Corporate Membership for contractors, consultants and suppliers
- Associate Membership for academic organizations.

The CBDG runs events, helps to produce information and has a number of Project Teams active, studying the following topics:

- publications
- new materials and techniques
- standards of workmanship and construction on site
- post-tensioning (1.4.1)
- concrete solutions for motorway widening
- education and training
- buildability.

1.7.6 Comité Euro-International du Béton (CEB)

CEB, the European Concrete Committee, is an international forum for the development of better use of concrete. From the work of its Subcommittees and Task Groups it produces Bulletins, including two which are relevant to bridge repair: Bulletin No. 148, *Durability of concrete structures*, 1982 and Bulletin No. 183, Design Guide, *Durability of concrete structures*, Thomas Telford, 1992. This document transposes the latest research findings into guidance for practising engineers. It explains the main transport mechanisms and degradation processes without going into full material science detail. Porosity to gases, water and dissolved agents depends upon pore size and distribution and cracks in the concrete.

CEB has a current Task Group 5/4: Assessment, maintenance and repair of concrete, with the following terms of reference.

- To develop rational and coordinated engineering approaches on technical and economic aspects of the assessment, maintenance and repair of existing structures.
- To develop Structures Management Systems related to inspection, maintenance and use. This will include reliability aspects of safety and serviceability.
- To develop rational techniques of protection, rehabilitation and strengthening (liaising with Commission 3).
- To utilize materials modelling (Task Group 5/1) on appropriate detailing level to forecast residual service life of structures depending on alternative choices of maintenance and repair.

- ☐ To provide alternative repair strategies to the main deterioration mechanisms defined by Task Group 5/1.
- ☐ To contribute to a coordinated approach together with other related international organizations.

Task Group 5/4 will adopt an engineering approach based on sound scientific macro-level modelling.

Testing and assessment work will be directed towards rational selection of test methods. Two strategies will be considered:

- ☐ statistical, requiring large scale sampling;
- ☐ selective, depending upon engineering experience.

Testing and monitoring should include performance of repairs. Work on repair techniques will focus on rational strategies for selecting optimum repair methods. Task Group 5/4 will continue cooperation with RILEM TC 124 on repair of corrosion-damaged structures. The Task Group recognizes that the purely technocratic solution of some problems may prove obsolete and sees the need for a multi-disciplinary approach, taking into account from the start economic and legal aspects.

1.7.7 Comité Européen de Normalisation (CEN)

The European standards organisation, CEN, was until 1988 a voluntary association of national standards bodies, including BSI. It is based in Brussels and partially funded by the EC. It covers all aspects of construction by means of a series of Technical Committees, each served by Subcommittees and each of these in turn served by Task Groups. Technical Committee TC104, deals with concrete with the help of a number of Subcommittees, one of which is TC104/SC8, Products and systems for the repair and protection of concrete, with the following Task Groups active:

TG1 Surface protection; coatings and impregnants.
TG2 Repair; cementitious mortars and polymer cements.
TG3 Structural bonding; external reinforcement.
TG4 Resin injection.
TG5 Anchoring or fixing.
TG7 Principles for repair.

TG6, Cathodic protection, reports to Subcommittee TC262/SC2 which deals with electrical systems. TC104 published document, ENV206: 1992 Concrete: Performance, production, placing and compliance criteria, which TC104/SC1 is reviewing and recasting with particular attention to the definition of deterioration mechanisms, an area covered recently by CEB in their Bulletin 183. TG2 is drafting clauses for repair on the basis of performance in accordance with ENV206: 1992. DAfSTb (1991) published guidelines for the protection and repair of concrete components.

When a Euronorm (EN), a CEN standard, is published it becomes mandatory as a model for a harmonized national standard which, in the UK, is issued by BSI. Type A Euronorms are design codes and Type B cover products. In all European Community countries Euronorms override any national standards which had previously been relevant.

1.7.8 Construction Industry Research and Information Association (CIRIA)

CIRIA is a non-profit-distributing, private sector organization carrying out research and providing information for its members, comprising all types of organizations concerned with construction. The following publications are relevant to the repair of concrete bridges.

Technical Note No. 130. *Protection of reinforced concrete by surface treatments.*
Technical Note No. 139. *Pull-off testing of repairs.*
Technical Note No. 143. *Testing concrete in structures.*

1.7.9 Concrete Society

The following publications are relevant to the repair of concrete bridges.

Technical Report No. 22. *Non-structural cracks in concrete*. 1992.
Technical Report No. 26. *Repair of concrete damaged by reinforcement corrosion*. 1984.
Technical Report No. 30. *Alkali-silica reaction. Minimising the risk of damage to con-*crete. Guidance notes and model specification clauses. 1987.
Technical Report No. 31. *Permeability testing of site concrete—a review of methods and* experience. 1988.
Technical Report No. 33. *Assessment and repair of fire-damaged concrete structures.* 1990.
Technical Report No. 35. *Underwater concreting.* 1990.
Technical Report No. 36. *Cathodic protection of reinforced concrete*. 1989.
Technical Report No. 37. *Model specification for cathodic protection of reinforced concrete*. 1991.
Technical Report No. 38. *Patch repair of reinforced concrete subject to reinforcement* corrosion. Model Specification and Method of Measurement. 1991.

The Concrete Society set up a Research Forum in 1992 whereby subscribers can propose research topics deserving support. Topics, which may be in the repair field, are entered in a database and considered by an Executive Committee to decide upon priorities.

1.7.10 Department of Transport (DoT) publications

Bridge Inspection Guide. HMSO. 1983.
Materials for the repair of concrete highway structures. Departmental Standard BD27/86.
Trunk road and motorway structures—records and inspection. Trunk Road Maintenance and Management Notice TRMM 2/88.
Impregnation of concrete highway structures. Departmental Advice Note BA33/90.
Inspection and repair of highway structures. Departmental Advice Note BA35/90.
Criteria and material for the impregnation of concrete highway structures. Departmental Standard BD43/90.

Fatigue of corroded reinforcement. Departmental Advice Note BA38/93.
Post-tensioned concrete bridges. Planning, organisation and methods for carrying out Special Inspections. Departmental Advice Note BA50/93.
TRL Laboratory Reports and other TRL publications are listed in the References to this chapter.

1.7.11 Fédération Internationale de la Précontrainte (FIP)

Guide to good practice—repair and strengthening of concrete structures. Thomas Telford Ltd.

1.7.12 International Association for Bridge and Structural Engineering (IABSE)

The International Association for Bridge and Structural Engineering is concerned with the design of new structures, including designing for durability. The Lisbon Conference 1989 addressed this aspect with the theme: Durability of concrete structures. The conference held sessions on:

- general durability aspects,
- material properties,
- design, detailing and construction,
- maintenance, repair, rehabilitation and alteration.

IABSE has a current Working Commission 8, Operation, maintenance and repair of structures, which is concentrating on practical aspects.

1.7.13 Institution of Civil Engineers (ICE)

The Institution's Infrastructure Policy Group (IPG) and the Association of Municipal Engineers (AME) promoted a debate which resulted in the Report (January 1988): The State of Roads and Bridges in the UK—the Case for Action. This paper was addressed to the political policy makers, civil engineers who advise them and the 130 local highway authorities throughout the UK. It reviewed the position since April 1986 when the IPG's Second Report noted deep concern about the state of roads and highway structures. Points concerning the repair of bridges were as follows.

(*a*) While the Government and some local authorities have established Bridge Management Systems, too many authorities have insufficient information systems.
(*b*) The state of some 11 260 out of about 54 000 mainly pre-1922 highway bridges under 50 m span warrants serious concern and substantial expenditure.
(*c*) Long-term deterioration can be disguised by short-term, low cost techniques and road users are largely unaware of the scale of structural deterioration.
(*d*) The picture of the State of Bridges provided by the Census and Survey omitted the Department of Transport (DoT), Welsh Office and Scottish

Office bridges, those over 50 m span and most of the shorter span bridges built since 1922.

The main thrust of the Report was the extent to which it was possible to tackle the problem and ways of raising finance for this purpose—matters outside the scope of this State-of-the-Art Review.

DoT (January 1987) published: *Assessment of Highway Bridges and Structures, Bridge Census and Sample Survey.* The ICE's comments on this document formed Annex A to the IPG/AME Report (1988). The ICE welcomed the study and saw the need for a structured programme to assess the total national bridge stock, to impose weight restrictions where necessary and to implement rehabilitations or renewals as appropriate.

DoT (November 1987) announced a 15-year rehabilitation programme for motorway and trunk road bridges. This was described by Holland and Dawe (1990) and comprises the following.

(i) Steady-state maintenance, comprising repair and replacement of deteriorated or damaged bridges.
(ii) Assessment and strengthening.
(iii) Upgrading substandard features, including waterproofing unprotected decks and rehabilitation of post-tensioned prestressed concrete bridges.

1.7.14 Institution of Highways and Transportation (IHT)

The National Workshops in 1986 and 1990 and the Symposia in 1992 and 1994 organized by IHT are listed in 1.7.2.

1.7.15 Institution of Structural Engineers (IStructE)

The IStructE provides administrative support to FIP and, jointly with the ICE, also to SCOSS. IStructE Working Parties reported on leading topics of the day, such as HAC concrete (1976) and ASR (1990 and 1992). A joint ICE/IStructE Working Party is considering the problems of post-tensioned tendons.

1.7.16 Specialist associations

Firms and individuals concerned with setting and maintaining high standards in their specialist fields have formed the following associations to advance their aims:

1969 Federation of Epoxy Resin Formulators and Applicators

1976 originally the Association of Gunite Contractors, now the Sprayed Concrete Association

1989 Concrete Repair Association

1992 Society for Cathodic Protection of Reinforced Concrete

1.7.17 Standing Committee on Structural Safety (SCOSS)

SCOSS is an independent body established in 1976 by the ICE, IStructE and others to maintain a continuing review of building and civil engineering matters affecting the safety of structures. It complements the existing administration for preserving structural safety and provides a route for the feedback of relevant information. It reports on topics of concern with recommendations for research or other action by appropriate organizations. Matters relevant to bridges include inspection, long-term performance of structural bonding, post-tensioning tendons and impact hazards.

1.8 References

AHLSKOG J. J. (1990). A. S. Nowak (ed.). Bridge management—the answer to the challenge. *Bridge evaluation, repair and rehabilitation.* Nato ASI Series E: Applied Sciences, Vol. 187, ISBN 0 7923 0999 5, 3–10.

ANDREWS G. (1984). Hundreds of bridges need costly repairs. *The Guardian.* 24 Aug. 1984, 2.

BARTON J. R. and KUSENBERGER F. N. (1981). Detection of defects in the reinforcement of prestressed concrete bridge members. *Proc. Int. Conf. on Inspection, Maintenance and Repair of Road and Railway Bridges*, 13–17 Apr., Bruxelles-Paris, 145–150.

BENNISON P. (1987). R. J. Cope (ed.). Repair and protection of reinforced concrete bridges. *Concrete bridge engineering: performance and advances*, ch. 4, Elsevier Applied Science.

BIJEN J. M. J. M. (1989). Maintenance and repair of concrete structures. *Heron*, **34**, No. 2, 3–33.

BLAKELOCK R. (1993). Experience in the development of a computerised bridge management system and of its use in a number of authorities in the UK and overseas. *Proc. 2nd Int. Conf. on Bridge Management.* University of Surrey, Guildford, 18–21 Apr.

BLIGHT G. E. (1990). Rehabilitation of reinforced concrete structures affected by alkali-silica reaction. *Structural Engineering Review*, **2**, No. 2, 113–120.

BLUNK G. and BRODERSEN H. A. (1980). Zum Widerstand von Beton gegenüber Harnstoff und Frost. *Strasse und Autobahn*, No. 3.

BOLLABAZ M. and HORNER R. M.W. (1990). Modelling and predicting bridge repair and maintenance costs. *Proc. 1st Int. Conf. on Bridge Management.* University of Surrey, Guildford, 28–30 Mar., 187–197.

BRIDGE INSPECTION PANEL (1983). *Bridge inspection guide.* Department of Transport, Scottish Development Department, Welsh Office and Department of the Environment for Northern Ireland. HMSO.

BRITISH CEMENT ASSOCIATION (1992). *Alkali-aggregate reaction 1977–1992.* A definitive bibliography.

BRITISH STANDARDS INSTITUTION (1984). *Steel, concrete and composite bridges.* BSI, London, BS 5400, Parts 1–10.

BRITISH STANDARDS INSTITUTION (1991). *Specification for salt for spreading on highways for winter maintenance.* BSI, London, BS 3247.

BROWN J. H. (1986). The performance of concrete in practice. *A field study of highway bridges.* TRRL Contractor Report 43.

BROWNE R. D. (1982). Design prediction of the life for reinforced concrete in marine and other chloride environments. *Durability of building materials: 1.* Elsevier, 113–125.

BUCKLAND P. G. (1990). Canada's advanced national standard on bridge evaluation. *Proc. 1st Int. Conf. on Bridge Management*. University of Surrey, Guildford, 28–30 Mar., 575–584.

BUENFELD N. (1992). Current projects. Private communication.

BUILDING RESEARCH ESTABLISHMENT (1982). *The durability of steel in concrete: Part 1 Mechanism of protection and corrosion*. Digest 263. HMSO.

BUNGAY J. A. (1992). Current projects. Private communication.

BURKE M. P. (1990). The integrated construction and conversion of single and multiple bridge spans. *Proc. 1st Int. Conf. on Bridge Management*. University of Surrey, Guildford, 28–30 Mar., 677–693.

CABRERA J. G., CUSENS A. R. and LYNSDALE C. J. (1989). Porosity and permeability as indicators of concrete performance. *Symposium on the Durability of Structures*. Lisbon. International Association for Bridge and Structural Engineering.

CAMOMILLA G., DRAGOTTI A., NEBBIA G. and ROMAGNOLO M. (1990). Programmed maintenance of motorway bridges: Italian experience in the use of expert systems. *Proc. 1st Int. Conf. on Bridge Management*. University of Surrey, Guildford, 28–30 Mar., 155–171.

CEB (1982). *Durability of concrete structures*. CEB Bulletin No. 148.

CEB (1992). *Durability of concrete structures*. CEB Bulletin No. 183.

CLARK L. A. (1990). Structural aspects of alkali-silica reaction. *Structural Engineering Review*, **2**, No. 2, 81–88.

CLARK L. A. (1992). *Performance in service of post-tensioned concrete bridges*. Published by BCA for CBDG, Oct., 59.

CLARK L. A. (1993). Assessment of concrete bridges with ASR. *Proc. 2nd Int. Conf. on Bridge Management*. University of Surrey, Guildford, 18–21 Apr.

COGSWELL G. and PALMER J. (1990). Bridging the age gap. *Proc. 1st Int. Conf. on Bridge Management*. University of Surrey, Guildford, 28–30 Mar., 24–26.

CONCRETE SOCIETY (1984). *Repair of concrete damaged by reinforcement corrosion*. Technical Report No. 26.

CONCRETE SOCIETY (1987). *Alkali-silica reaction*. Minimising the risk of damage to concrete. Guidance notes and model specification clauses. Technical Report No. 30.

CONCRETE SOCIETY (1988). *Permeability testing of site concrete—a review of methods and experience*. Technical Report No. 31.

CONCRETE SOCIETY (1989(a)). *Concrete bridges—management, maintenance and renovation*. Proceedings of a one-day conference held in London, 23 Feb., CS. 102, 40.

CONCRETE SOCIETY (1989(b)). *Cathodic protection of reinforced concrete*. Technical Report No. 36.

CONCRETE SOCIETY (1990). *Assessment and repair of fire-damaged concrete structures*. Technical Report No. 33.

CONCRETE SOCIETY (1990). *Underwater concreting*. Technical Report No. 35.

CONCRETE SOCIETY (1991). *Model specification for cathodic protection of reinforced concrete*. Technical Report No. 37.

CONCRETE SOCIETY (1991). *Patch repair of reinforced concrete subject to reinforcement corrosion*. Model specification and method of measurement. Technical Report No. 38.

CONCRETE SOCIETY (1992). *Non-structural cracks in concrete*. Technical Report No. 22.

COPE R. J. (ed.). (1987). *Concrete bridge engineering—performance and advances*. Elsevier Applied Science, 1987.

COPE R. J. and SLADE L. (1990). The shear capacity of reinforced concrete members subjected to alkali-silica reaction. *Structural Engineering Review*, **2**, No. 2, 105–112.

DAfStb (1991). *Guidelines for the protection and repair of concrete components*. German Committee on Reinforced Concrete. Section VII of the Standardization Committee for

Construction.

DAGHER H., HAMILTON W. and KULENDRAN S. (1990). A. S. Nowak (ed.). Durability of concrete bridge decks with reduced reinforcement. *Bridge evaluation, repair and rehabilitation.* Nato ASI Series E: Applied Sciences, Vol. 187, ISBN 0 7923 0999 5.

DAHINTER K. (1990). Inspection and strength evaluation of concrete highway bridges in Czechoslovakia. *Proc. 1st Int. Conf. on Bridge Management.* University of Surrey, Guildford, 28–30 Mar., 383–393.

DEPARTMENT of the ENVIRONMENT (1974). *High alumina cement concrete.* Technical Memorandum BE5/74.

DEPARTMENT of INDUSTRY (1981). *The corrosion of steel, and its monitoring, in concrete.* Guide to practice in corrosion control No. 7.

DEPARTMENT of TRANSPORT (1977). *The inspection of highway structures.* Technical Memorandum BE4/77.

DEPARTMENT of TRANSPORT (1984). *Assessment of highway bridges and structures.* Departmental Standard BD21/84.

DEPARTMENT of TRANSPORT (1986). *Materials for the repair of concrete highway structures.* Departmental Standard BD27/86.

DEPARTMENT of TRANSPORT (1987). *Assessment of highway bridges and structures, bridge census and sample survey.*

DEPARTMENT of TRANSPORT (1988). *Trunk road and motorway structures—records and inspection.* Trunk road maintenance and management notice, TRMM 2/88.

DEPARTMENT of TRANSPORT (1990). *Impregnation of concrete highway structures.* Departmental Advice Note BA33/90.

DEPARTMENT of TRANSPORT (1990). *Inspection and repair of highway structures.* Departmental Advice Note BA35/90.

DEPARTMENT of TRANSPORT (1990). *Criteria and material for the impregnation of concrete highway structures.* Departmental Standard BD43/90.

DEPARTMENT of TRANSPORT (1993). *Fatigue of corroded reinforcement.* Departmental Advice Note BA38/93.

DEPARTMENT of TRANSPORT (1993). *Post-tensioned concrete bridges. Planning, organisation and methods for carrying out Special Inspections.* Departmental Advice Note BA50/93.

DEPARTMENT of TRANSPORT (1993). Private communication.

ENGINEERING NEWS RECORD (1977). One in six US highway bridges is deficient. *Engineering News Record*, 10 Mar. 1977, 18–21.

EVERETT L. H. and TREADAWAY K. W. J. (1971). Effect of reinforcement corrosion. Durability of concrete. *Supplement to The Consulting Engineer*, Apr./May.

FEDERAL HIGHWAY ADMINISTRAION (1970). *Bridge inspector's training manual 70.* US Department of Transportation, Bureau of Public Roads, Washington DC.

FIP (1991). Repair and strengthening of concrete structures. *Guide to good practice.* Thomas Telford, London.

FREARSON J. P. H. (1985). The problem—identification and interpretation. *Concrete bridges—investigation, maintenance and repair.* London Tara Hotel. 25 Sept.

GLEDHILL C. and JONES C. J. F. P. (1981). The repair of bridges damaged by vehicular impact. *Proc. Int. Conf. on Inspection, Maintenance and Repair of Road and Railway Bridges*, 13–17 Apr., Bruxelles-Paris, 339–344.

GODDARD J. (1989). Bridging the concrete gap. *Surveyor*, 6 Apr.

HALSTEAD P. E. (1971). Causes of concrete decay. Durability of concrete. *Supplement to The Consulting Engineer*, Apr./May.

HOBBS D. W. (1990). Cracking and expansion due to alkali-silica reaction: its effect on concrete. *Structural Engineering Review*, **2**, No. 2, 65–80.

HOLDEN W. R., PAGE C. L. and SHORT N. R. (1983). The influence of chlorides and sulphates on durability. Presented at meeting: *Corrosion of reinforcement in concrete construction*, London, 13–15 June. Society of Chemical Industry, 143–150.

HOLLAND D. A. and DAWE P. H. (1990). Bridge rehabilitation: Department of Transport's fifteen-year strategy. *Proc. 1st Int. Conf. on Bridge Management*. University of Surrey, Guildford, 28–30 Mar., 135–144.

IDORN G. M. (1967). *Durability of structures in Denmark*. Thesis for DSc. Technical University of Denmark, Copenhagen.

INGVARSSON H. (1988). Operation and maintenance of bridges. Pre-conference proceedings: *Towards better road performance*. Riyadh, Saudi Arabia, 13–18 Jan., 6.31–6.58.

INGVARSSON H. (1990). Bridge operation and maintenance costs. *Proc. 1st Int. Conf. on Bridge Management*. University of Surrey, Guildford, 28–30 Mar., 199–204.

INSTITUTION OF CIVIL ENGINEERS (1988). *The state of roads and bridges in the UK—the case for action*. Infrastructure Policy Group (IPG) and the Association of Municipal Engineers (AME).

INSTITUTION OF STRUCTURAL ENGINEERS (1976). Report of Working Party on high alumina cement concrete. *The Structural Engineer*, Sept., **54**, No. 9, 352–361.

INSTITUTION OF STRUCTURAL ENGINEERS (1990). The structural effects of ASR. *Structural Engineering Review Special Issue*, **2**, No. 2, June.

JOHN D. G. (1992). Current projects. Private communication.

KAY T. (1992). *Assessment and renovation of concrete structures*. Concrete Design and Construction Series. Longman Scientific and Technical.

KLOPFER H. (1981). The carbonation of external concrete and how to combat it. *One-day conference on the repair of concrete structures*. Imperial College, London.

KREPS-HEYNDRIKX Y. and CALLENS M. (1981). Travaux de réparation d'un pont précontraint, heurté par un véhicule. *Proc. Int. Conf. on Inspection, Maintenance and Repair of Road and Railway Bridges*. 13–17 Apr., Bruxelles-Paris, 366–370.

KULASH D. J. (1990). SHRP's concrete and structures research: goals and recent developments. VTI Rapport 1990 352A, *Proc. of Strategic Highways Research Program and Traffic Safety on two Continents*, Gothenburg, Sweden, 27–29 Sept., 69–82.

KEER J. G. and LE PAGE B. H. (1990). Crack bridging by surface treatments to concrete. *Proc. Int. Conf. on Bridge Management*. University of Surrey, Guildford, 28–30 Mar., 233–241.

LEEMING M. B. (1990). Keeping water out of concrete—the key to durability. *Proc. Int. Conf. on Bridge Management*. University of Surrey, Guildford, 28–30 Mar., 243–258.

LOE J. A., IRWIN C. A. K. and GRAINGER J. W. (1975). *Loading tests on three floors containing high alumina cement joists*. TRRL Supplementary Report 172UC.

MAILVAGANAM N. P. (1992). *Repair and protection of concrete structures*. Thomas Telford, 1992, ISBN 0 8493 4993 1.

MALHOTRA H. L. (1984). *Spalling of concrete in fires*. CIRIA Technical Report No. 118.

MANSON J. A. *et al.* and CADY P. D. *et al.* (1983). *Long-term rehabilitation of salt-contaminated bridge decks*. NCHRP Report 257. Transportation Research Board.

MARTIN H., RAUSEN A. and SCHIESSL P. (1975). Carbonation of concrete made with different types of cement. *Behaviour in service of concrete structures*. FIP/CEB/RILEM Colloquium, Liege.

MAYS G. C. (ed.). (1992). *Durability of concrete structures. Investigation, repair, protection*. RMCS (Cranfield), UK, E. and F. N. Spon, London, ISBN 0 419 15620 8, 1992.

MAYS G. C. (1992). Current projects. Private communication.

MAYS G. C. and BARNES R. A. (1991). The performance of lightweight aggregate concrete structures in service. *The Structural Engineer*. 15 Oct., **69**, No. 20, 351–361.

MIYAGAWA T. (1991). Durability design and repair of concrete structures: chloride corrosion of reinforcing steel and alkali-silica reaction. *Mag. Concr. Res.*, **43**, No. 156, 155–170.

MIYAGAWA T., HISADA M., INOUE S. and FUJII M. (1991). Effect of concrete surface treatment on expansion due to alkali-silica reaction. *Concrete Library of JSCE*, No. 18, Dec., 237–261.

MINISTERE DE L'EQUIPEMENT (1975). *Défauts apparents des ouvrages d'art en béton*. SETRA and LCPC, Paris.

NEVILLE A. and WAINWRIGHT P. J. (1975). *High alumina cement concrete*. Construction Press, Lancaster.

NEVILLE A. (1983). Corrosion of reinforcement. *Concrete*, June, **17**, No. 6, 48–49.

NEVILLE A. (1985). What worries me ... *Concrete Technology*, 7 Mar.

NEW CIVIL ENGINEER (1983). Swiss work to beat bridge decay. *New Civil Engineer*, 17 Nov. 1983, 26–30.

NEW CIVIL ENGINEER (1984). Corrosion threatens modern bridges. *New Civil Engineer*, 23–30 Aug. 1984, 4.

NEW CIVIL ENGINEER (1985). Welsh bridge failure starts tendon scare. *New Civil Engineer*, 12 Dec. 1985, 4, 50.

NEW CIVIL ENGINEER (1986). Symptoms of sulphates. *New Civil Engineer*, 10 July 1986, 16–17.

NEW CIVIL ENGINEER (1993). Long term problems for HAC. *New Civil Engineer*, 11 Feb. 1993, 4.

NIXON P. J. (1985). Changes in the properties of OPC. *Concrete Technology*, 7 Mar.

OECD (1972). *Waterproofing of concrete bridge decks*. Road Research Group Report. Organisation for Economic Co-operation and Development, 2, rue André Pascal, 75775 Paris Cedex 16.

OECD (1976). *Bridge inspection*. Road Research Group Report. Organisation for Economic Co-operation and Development, 2, rue André Pascal, 75775 Paris Cedex 16.

OECD (1979). *Evaluation of load carrying capacity of bridges*. Road Research Group Report. Organisation for Economic Co-operation and Development, 2, rue André Pascal, 75775 Paris Cedex 16.

OECD (1981). *Bridge maintenance*. Road Research Group Report. Organisation for Economic Co-operation and Development, 2, rue André Pascal, 75775 Paris Cedex 16.

OECD (1983). *Bridge rehabilitation and strengthening*. Road Transport Research Group Report. Organisation for Economic Co-operation and Development, 2, rue André Pascal, 75775 Paris Cedex 16.

OECD (1989). *Durability of concrete bridges*. Road Transport Research Group Report. Organisation for Economic Co-operation and Development, 2, rue André Pascal, 75775 Paris Cedex 16.

OECD (1992). *Bridge management*. Scientific Expert Group Report. Organisation for Economic Co-operation and Development, 2, rue André Pascal, 75775 Paris Cedex 16.

ONO K. (1990). Strength and stiffness of alkali-silica reaction concrete and concrete members. *Structural Engineering Review*, **2**, No. 2, 121–126.

OWENS P. L. (1985). Illusions of durability—a specifier's nightmare. *Concrete Technology*, 7 Mar.

PAGE C. L. (1981). *Corrosion of steel in concrete*. One-day conference on the failure and repair of corroded reinforced concrete structures. Europa Hotel, London, 6 Nov., 1–16.

PAGE C. L. and TREADAWAY K. W. J. (1982). *Nature*, 297, 5862, 109–115.

PAGE C. L. and LAMBERT P. (1986). *Analytical and electrochemical investigations of reinforcement corrosion*. TRRL Contractor Report CR30.

PAGE C. L. (1987). An overview of current research. *Proc. 2nd Int. Conf. Deterioration and Repair of Reinforced Concrete in the Arabian Gulf*, **2**, Discussion, 161–171.

PAGE C. L., TREADAWAY K. W. J. and BAMFORTH P. B. (1990). *3rd Int. Symp. on Corrosion of Reinforcement in Concrete*. Wishaw, Warwickshire. SCI. Elsevier Applied Science. Papers 21–24 May.

PAGE C. L., SERGI G. and THOMPSON D. M. (1991). Electrochemical induction of alkali-silica reaction in concrete. *Materials and Structures*, **24**, 359–361.

PAGE C. L. (1992). Current projects. Private communication.

PERKINS P. H. (1986). *Repair, protection and waterproofing of concrete structures*. Elsevier Applied Science.

PETRANGELI M. P. (1990). Inspection and repair of some bridges in Italy. *Proc. 1st Int. Conf. on Bridge Management*. University of Surrey, Guildford, 28–30 Mar., 373–381.

POCOCK D. C. and GIBBS B. (1986). Diagnosis, prevention and repair of deterioration in concrete bridges—an overview. *Sino-British Highways and Urban Traffic Conf.* Institution of Highways and Transportation. Beijing Papers.

PRICE A. R. (1989). *A field trial of waterproofing systems for concrete bridge decks*. TRRL Laboratory Report 185.

PULLAR-STRECKER P. (1987). Corrosion-damaged concrete. *Assessment and repair*. CIRIA and Butterworths and Co. Ltd, London.

PULLAR-STRECKER P. (1988). Corrosion-damaged concrete. *Concrete Repair Conference*, London Tara Hotel, 16 Mar., Palladian Publications Ltd.

ROBERTS M. H. (1981). *Carbonation of concrete made with dense natural aggregates*. Building Research Establishment Information Paper 6/81.

ROSTAM S. (1989). Management and maintenance of concrete bridges. An assessment of the European scene. *Construction and Building Materials*. Sept., **3**, No. 3, 159–163.

SADEGZADEH M. and PAGE C. L. (1990). *Effects of urea on the durability of reinforced concrete*. TRRL Contractor Report 208.

SERGI G., PAGE C. L. and THOMPSON D. M. (1991). Electrochemical induction of alkali-silica reaction in concrete. *Materials and Structures*, **24**, 359–361.

SHACKLOCK B. W. (1971). Admixtures and concrete durability. Durability of concrete. *Supplement to The Consulting Engineer*, Apr./May.

SIEMES T. and VROUWENVELDER T. (1985). Durability—a probabilistic approach. *International Association for Bridge and Structural Engineering Proceedings*, P-85/85, 65–76.

SMITH N. J. (1990). Management of bridgeworks maintenance in the UK. *Proc. 1st Int. Conf. on Bridge Management*. University of Surrey, Guildford, 28–30 Mar., 223–230.

STEVENS R. F. (1974). *Suggestions for testing high alumina cement concrete*. 17 July, Structural Properties Division, Building Research Establishment.

SWAMY R. N. (1990). Alkali-silica reaction and concrete structures. *Structural Engineering Review*, **2**, No. 2, 89–104.

THOMAS M. D. A. (1991). Marine performance of PFA concrete. *Mag. Conc. Res.*, **43**, No. 156, 171–185.

THOMAS M. D. A. and MATTHEWS J. D. (1991). *Durability studies of PFA concrete structures*. BRE Information Paper IP 11/91.

THOMPSON D. M. (1992). Current projects. Private communication.

TILLY G. P. (1985). Developments in research to meet problems. *Concrete bridges: investigation, maintenance and repair*. Proc. of a one-day Symposium. London Tara Hotel, 25 Sept.

VAN DER TORN A. and REIJ A. W. F. (1990). A systematic approach to future maintenance. *Proc. 1st Int. Conf. on Bridge Management*. University of Surrey, Guildford, 28–30 Mar., 215–222.

VASSIE P. (1984). Reinforcement corrosion and the durability of concrete bridges. Paper 8798. *Proc. Instn Civ. Engrs*, Part 1, 1984, **76**, Aug., 713–723.

VASSIE P. (1985). Reinforcement corrosion and the durability of concrete bridges. 8798 Discussion. *Proc. Instn Civ. Engrs*, Part 1, 1985, **78**, Oct., 1253–1256.

VASSIE P. R. (1993). Secondary effects of ASR on bridges—corrosion of reinforcement steel. *Proc. 2nd Int. Conf. on Bridge Management*. University of Surrey, Guildford, 18–21 Apr.

WALLBANK E. J. (1989). *The performance of concrete in bridges—a survey of 200 highway bridges*. HMSO, London.

WILKINS N. J. M. and LAWRENCE P. F. (1983). The corrosion of steel reinforcements in concrete immersed in seawater. Presented at meeting: *Corrosion of reinforcement in concrete construction*. London, 13–15 June, Society of Chemical Industry, 119–141.

WOOD J. G. M. (1985). AAR—the growing problem. *Concrete Technology*, 7 Mar.

WOOD J. G. M., JOHNSON R. A. and ELLINAS C. (1990). Reliability analysis applied to deteriorating bridge structures. *Proc. 1st Int. Conf. on Bridge Management*. University of Surrey, Guildford, 28–30 Mar., 501–513.

WOOD J. G. M. (1990). Predicting future decay in concrete structures. *The design life of structures*. G. Somerville (ed.), British Group of IABSE Colloquium, Cambridge, July, Blackie Academic and Professional (an imprint of Chapman and Hall). Glasgow, 1992.

WOODWARD R. J. (1981). *Conditions within ducts in post-tensioned prestressed concrete bridges*. TRRL Laboratory Report 980.

WOODWARD R. J. (1981). *Case studies of the corrosion of reinforcement in concrete structures*. TRRL Laboratory Report 981.

WOODWARD R. J. and LOE J. A. (1981). The inspection and maintenance of post-tensioned concrete bridges. *Proc. Int. Conf. on Inspection, Maintenance and Repair of Road and Railway Bridges*. 13–17 Apr., Bruxelles-Paris, 265–270.

WOODWARD R. J. and WILLIAMS F. W. (1988). Collapse of Ynys-y-Gwas Bridge, West Glamorgan. Paper 9357. *Proc. Instn Civ. Engrs*, Part 1, 1988, Aug., 635–670.

WOODWARD R. J. (1989). *Measurements of prestress and load tests to failure on segmental concrete beams*. TRRL Research Report 255.

WOODWARD R. J. (1991). Deformation of segmental post-tensioned bridges as a result of corrosion of the tendons. *Proc. Instn Civ. Engrs*, Part 1, 1991, **90**, Apr., 397–419.

2

Case studies

2.1 Introduction

The following summaries illustrate the variety of problems met and solutions found in the rehabilitation of concrete bridges. They are grouped broadly according to the dominant location or cause of damage to facilitate comparisons, like with like, but they have no statistical status. It is mainly a collection of some of the cases that happen to have been published.

2.2 Superstructure

2.2.1 Flixton Estate Bridge, Homersfield — Stanley (1979), *The Guardian* (1991) and McCarter (1993)

Flixton Estate Bridge spans 16.5 m over the River Waveney on the Norfolk–Suffolk border near Bungay. It was commissioned in 1869 by Sir Shafto Adair, Bart., from Henry Eyton of Ipswich to replace an existing bridge on his Flixton Estate. Designed for a distributed load of 200 t to Eyton's requirements by W. and Thos. Phillips of the Iron Depot, Coal Exchange, London, it was built in the early 1870s for £344. Stanley (1979) recorded that embedding the iron frame in concrete was in accordance with the 'Patent Fireproof Specification of T. and W. Phillips'.

Two variable depth wrought iron trusses 3.7 m apart are connected at their curved bottom angles by wrought iron I sections at 600 mm centres. Between these cross-members are smaller wrought iron I sections in a herringbone pattern at 150 mm centres. This frame supports in-situ concrete which was poured to the depth of the trusses. Decorative precast cement-rich mortar panels were attached to the fascias. The rest of the fascias and the soffit were rendered with cement mortar.

It is privately owned and carried public traffic on the B1062 road until 1970, when a new road bridge was built and the old bridge was retained for pedestrians. Inspections found voided in-situ concrete of poor quality and badly corroded ironwork which had forced off the rendering in places. *The Guardian* (1991) reported plans to raise £50 000 to restore the oldest iron-reinforced bridge in Britain. In view of the deterioration of this Grade II* listed struc-

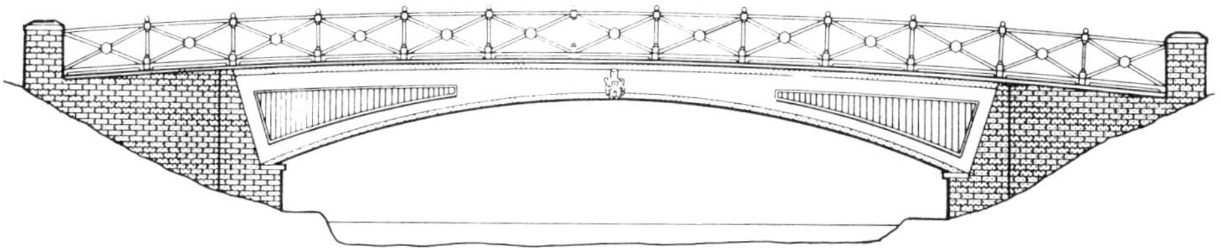

Fig. 2.1. Flixton Estate Bridge, Homersfield (Norfolk County Surveyor, 1993)

ture, the Norfolk County Surveyor (1993) designed rehabilitation works to restore it as shown in Figure 2.1.

Exposed ironwork will be shot-blasted and protected with three coats of zinc-rich paint. The ornate cast iron open parapet will be refurbished and the decorative panels of rendering bonded with epoxy resin or repaired with high-build polymer modified cementitious repair mortar. Significant cracks will be injected with low viscosity epoxy resin and a nominal 15 mm thick rendering applied to the soffit. The bridge will also be waterproofed, reinstated with a lightweight aggregate no fines concrete and resurfaced. The work is to be carried out by the new joint owners, Norfolk and Suffolk Historic Buildings Trusts.

2.2.2 Strengthening a 50-year old viaduct—Grozier (1974)

The viaduct over the River Cam flood plain comprises a river span of 13 m and eight spans of 7 m. It was built in 1926 to carry a railway to a cement works. After 47 years without any significant maintenance, the bridge was assessed in view of the 34 t wagons being replaced by 50 t wagons. There was some corrosion and spalling but only hairline cracks and no evidence of distress from the existing loading. Each span was strengthened by means of four 25 mm Macalloy bars stressed against steel brackets secured by similar bars passing through holes carefully drilled in the haunches of each beam. After removing defective concrete, the surfaces were grit blasted, wrapped with 50 mm galvanized mesh and gunited with 7:2 sand:rapid hardening cement delivered at about 0.5 N/mm^2. Water added at the nozzle gave water/cement ratio 0.25 to 0.30 and compressive strength of about 70 N/mm^2. The stresses in the repaired and strengthened structure were lower under the new loading than previously.

2.2.3 Royal Tweed Bridge, Berwick—Estie (1993)

Built in 1928, the overall length of 428 m includes four open spandrel arches of 51, 75, 87 and 110 m. After more than 50 years, inspection found signs of deterioration. Although the carbonation depths were only 3 to 4 mm, the marine environment and leakage of deicing salt through joints had caused cracking and spalling, especially in parts of the deck slab and beams where the concrete was porous or honeycombed. Rehabilitation in 1980 included vacuum grouting with epoxy resin, trimming the spalled areas, grit blasting the concrete substrate and reinforcement and making **good** with several repair mortars.

2.2.4 Haddiscoe Bridge—Cavalier and Vassie (1981 and 1982)

Constructed in Norfolk in 1960, each of the thirteen simply supported spans has seven universal steel beams with composite 190 mm reinforced concrete deck slab. Cover to reinforcement was 25 mm nominally and the concrete surface was treated with two coats of bitumen emulsion but no effective waterproofing.

In 1978 work on expansion joints and resurfacing revealed some local disintegration of the concrete and reinforcement pitting corrosion. The greater part of the deck slab appeared to be sound but stained areas could be broken out to show corrosion of reinforcement. Thorough investigation determined:

- chloride penetration with high probability of corrosion,
- half-cell potentials more negative than -250 mV,
- concrete cover ranged from 20 to 35 mm,
- carbonation depths were less than 5 mm,
- resistivity substantially below 5000 Ωcm in regions where reinforcement was badly corroded.

The extent of repairs was decided in the light of all of these observations. On the three central spans 52% of the area had more than 0.7% chloride ion by weight of cement and was broken out to a depth of 75 mm; a further 8% had 0.4 to 0.7% chloride and was broken out to 25 mm deep. On the approach spans 20% was broken out to 75 mm and a further 12% to 25 mm and the repaired deck waterproofed with two layers of mastic asphalt.

After more than 10 years, the 75 mm repairs had performed well with no further corrosion due to chloride but 25 mm repairs had tended to delaminate and required further work. Corrosion of steel in the original concrete contiguous with a repair was found in 1 out of 13 test areas.

2.2.5 D-T-D Canal Bridge, former Yugoslavia—Pakvor, Tarana and Darijevic (1983)

Inspection of the reinforced concrete arch road bridge over the Danube–Tisza–Danube Canal at Srpski Miletic discovered cracks in suspenders. Repairs were carried out without interrupting traffic, except for trial loading and testing which proved the repair successful.

2.2.6 Zuoz Bridge—Favre (1969), Billington (1979) and Angerer (1993)

The Zuoz bridge, illustrated in Figure 2.2, was built by Robert Maillart in the Swiss canton of Graubünden in 1901. It is a three hinged arch of twin hollow box section spanning 38 m over the River Inn.

Billington (1979) described contraction cracks which appeared in 1903 on the sunny side of the bridge. Neither Maillart nor anyone else considered that the

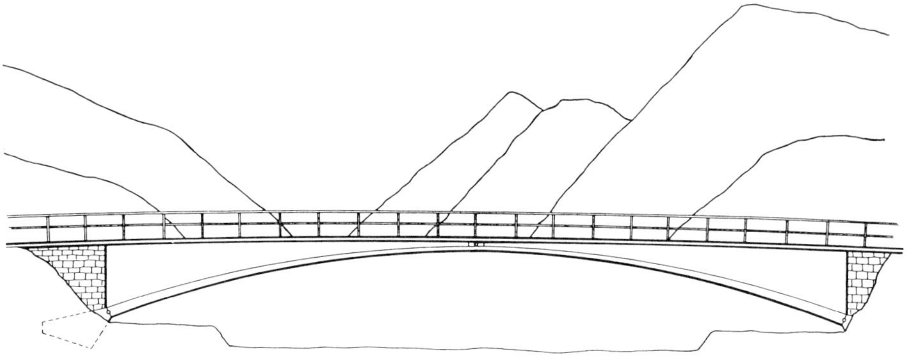

Fig. 2.2. Zuoz Bridge, Switzerland (Billington, 1979)

cracks were of any structural significance and Maillart suggested that the side walls could be whitewashed to improve their appearance. In 1905 his design for Tavanasa omitted concrete from the fascia areas which had cracked at Zuoz, material which carried little load but caused a visual problem. Billington gave full references and notes.

Favre (1969) described rehabilitation work as both speedy and cheap (total cost including fees SFr 85 000, that is barely 400 SFr/m^2) as follows:

- remove about 20 cm of hard Macadam;
- demolish the deck slab, inspect and make good the interior of hollow boxes as necessary;
- shutter, fix reinforcement and concrete new deck slab;
- roughen and gunite the bridge fascias and soffit;
- stabilize each abutment with four 50 t ground anchors 15 m long.

Angerer (1993) confirmed that no significant repairs have been required since 1969; only resurfacing the roadway with 7 cm of Gussasphalt in 1987. (See also 6.1.1.)

2.2.7 Fürstenland Bridge, St Gallen—Schneppendahl (1984)

This major open spandrel arch bridge was built in 1939–40 and, after 40 years, showed signs of deterioration, mainly on the underside and outer kerb of the cantilevered pedestrian deck. The work was carried out from two platforms, each 6 m long and suspended from a mobile gantry which did not obstruct mopeds, cyclists or pedestrians, while the traffic on the carriageway was completely unaffected. All unsound concrete was removed and the concrete substrate and exposed reinforcement prepared by grit and water blasting. Two coats of an anti-corrosion product were applied to the cleaned reinforcement and patch repairs completed using polymer-modified cement based mortar. The outer kerb of the pedestrian deck was treated with two layers of polymer-modified cementitious high-build coating reinforced with glass fibre alkali-resistant fabric. Finally, the underside of the pedestrian deck was given two coats of an emulsion based on acrylic copolymers to retard carbonation.

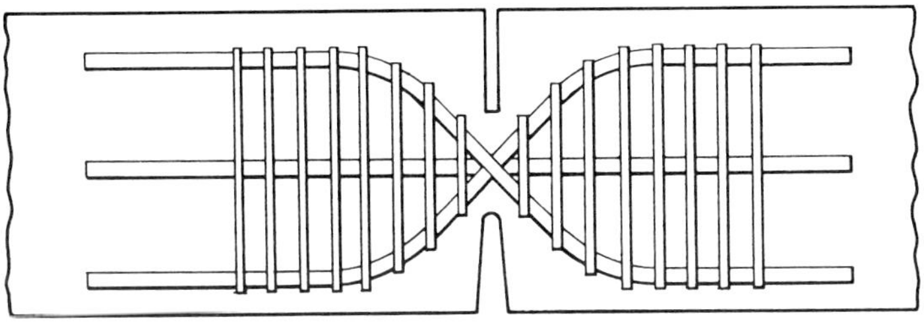

Fig. 2.3. Reinforced concrete hinge (New Civil Engineer, 1987, and DoT, 1993)

2.2.8 Monitoring hinges—*New Civil Engineer* (1987) and Department of Transport (1993)

Constructed in the early 1960s, the 60° skew, 146 m long reinforced concrete slab bridge carries a minor road over the M1 motorway. Approach spans cantilevered over raking circular column supports and a short balanced cantilever section over a pier in the central reservation of the motorway support suspended spans over each carriageway by means of reinforced concrete hinges as shown in Figure 2.3.

These hinges run parallel to the motorway across most of the width of the bridge but, near the edges, perpendicular to the axis of the bridge. This articulation caused a concentration of stresses due to live loading, thermal and creep effects. Bars exposed at selected locations were found to be in good condition and residual stresses were checked by a specialist drilling technique. (See 3.5.)

2.2.9 Manama–Sutra Causeway bridges—Frearson *et al.*, Marshall and Channa (1987)

The bridges in the Bahrain causeway were built between 1974 and 1976. Frearson, Marshall and Hamilton (1987) found the extensive cracking, corrosion and lamination was due to air borne chloride while the carbonation had not reached the reinforcement. Marshall and Channa (1987) discussed various repair options and, after a full technical and economic appraisal, it was decided to replace the deck slabs on the steel beams.

2.2.10 M6 widening: Thelwall Viaduct—*New Civil Engineer* (1988)

Having carried up to 134 000 vehicles a day over the Manchester Ship Canal, the 1.3 km viaduct will be retained for northbound traffic when a new viaduct has been completed for southbound traffic. Corrosion of the deck slab reinforcement resulted in spalling and potholes in the carriageway. Cores revealed chloride concentrations up to 2.6% by weight of cement which required the entire deck to be replaced.

2.2.11 Barton High Level Bridge—*New Civil Engineer* (1986) and Jackson (1989)

In the course of upgrading the M63 from dual two-lane to dual three-lane, this 740 m structure built in 1959 had to be widened. Deterioration of the concrete and the reinforcement was so widespread that complete replacement of the deck was more economic than patch repairs. Demolition was carried out with a Whiphammer supplied by Wolfverine Technology, USA. Traffic management governed the three-phase construction sequence and the work was completed in 1989.

2.2.12 Punching shear failure of bridge decks—Al-Mandil, Azad, Baluch, Pearson-Kirk and Sharif (1989)

A survey conducted in a National Project on the condition of bridge decks in Saudi Arabia revealed more than 120 decks with varying degrees of soffit cracking. Grid patterned cracking on the soffits was diagnosed as flexural, caused and propagated by heavy vehicular loading, apparently by fatigue action, and leading eventually to local punching shear failures.

A bridge near Rumah, some 150 km north-east of Riyadh, was closely investigated by site tests, theoretical analysis and laboratory testing of panels. Although the uncracked deck slabs had sufficient punching shear strength, localized weaknesses were caused by the nucleation of flexural cracks. No repair method is described.

2.2.13 Repair of the Veauche Bridge—Archier (1989)

The relatively unusual reinforced concrete structure built in the 1930s is 165 m long and 7 m wide and comprises encastré arches with intermediate suspended spans. Deterioration was due to two main causes:

- the load carrying capacity of the structure designed to 1927 requirements was limited when compared with the 1971 standards;
- the effects of ageing and construction faults.

This necessitated the rehabilitation of the whole structure, which was carried out in three main phases. Almost all of the work was underneath the structure, using two different mobile platforms and scaffolding.

- Cleaning and preparing concrete surfaces, constructing new diaphragms capable of supporting anchorages for the prestressing tendons, injection of live cracks with an elastic material and stabilized cracks with a non-solvent resin having high adhesive strength. This had to be done in six weeks while traffic was prohibited.
- Work preparatory to prestressing, installing anchorages and the tendons in galvanized steel 101 mm diameter sheaths 2 mm thick with ends fitted together and sealed with epoxy resin.
- Applying prestress with the road closed to traffic, then injecting the sheaths with a specially formulated protective grout and completing all the other concrete rehabilitation. This included sand blasting with silica-free abrasive, coating exposed bright steel immediately with epoxy zinc paint, patch repairs using pure epoxy mortar, hydraulic mortar or resin paste, depending on the nature and location of the repair and sprayed concrete. The constituents for sprayed concrete were 300 kg cement, 1200 kg sand

0–5 mm and 600 kg grit 3–8 mm with water delivered to the nozzle at 0.7 N/mm^2 pressure giving a water/cement ratio of 0.5. It was applied in two passes; the first one of 15 mm to restore adequate cover and the second one up to 5 mm to provide a uniform appearance.

2.2.14 Three German bridges—Zichner (1989 and 1993)

Shortcomings in design or construction, accidents and changes of use lead to the need for rehabilitation which can be combined with strengthening. This may deliver a more robust structure which requires less maintenance than the minimum strength solution. On the other hand, there may be constraints which lead to a decision to demolish and rebuild. Three cases illustrated the variety of problems met.

(1) Saar Bridge

Built in 1955, the 147 m long structure comprised three two-hinged arches, each with three ribs, deck slab and crossbeams, post-tensioned longitudinally and transversely. During routine inspection, cracks were discovered near the crown of all three arches. Crack widths in the small land span were 0.2–0.3 mm, in the middle arch 1.2 mm and in the large river span 2.5 mm.

Prestressing tendons showed corrosion damage, some were fully stressed, some partially and some were unstressed and mortar in the ducts was found to be contaminated with chloride, cyanide, sulphide, sulphate and phosphate. Rehabilitation would have required such costly external prestressing and measures to prevent further cracking and protect against corrosion that the case was made to demolish and rebuild.

The bridge was demolished in 1986.

(2) Mosel Bridge

Built in 1963, the four prestressed concrete spans were as shown in Figure 2.4(a) with the short northern span anchored to prevent uplift.

At the beginning of the 1980s, the bridge showed two severe defects:

- ☐ a heavily corroded and immovable anchorage link and
- ☐ deflection of the hinge at the centre of the main span.

These defects adversely affected stability and serviceability. The kink in the river span was unsightly and affected driver comfort. The following work was completed in 1986.

- ☐ The deck was lifted to its correct level.
- ☐ Bearings were replaced.
- ☐ The centre hinge joint was closed.
- ☐ Supplementary prestressing installed as Figure 2.4(b) and (c).
- ☐ Defective anchorage no longer required so removed.
- ☐ Concrete ballast placed in hollow box north span.
- ☐ Waterproofing, surface finishes, etc. applied.

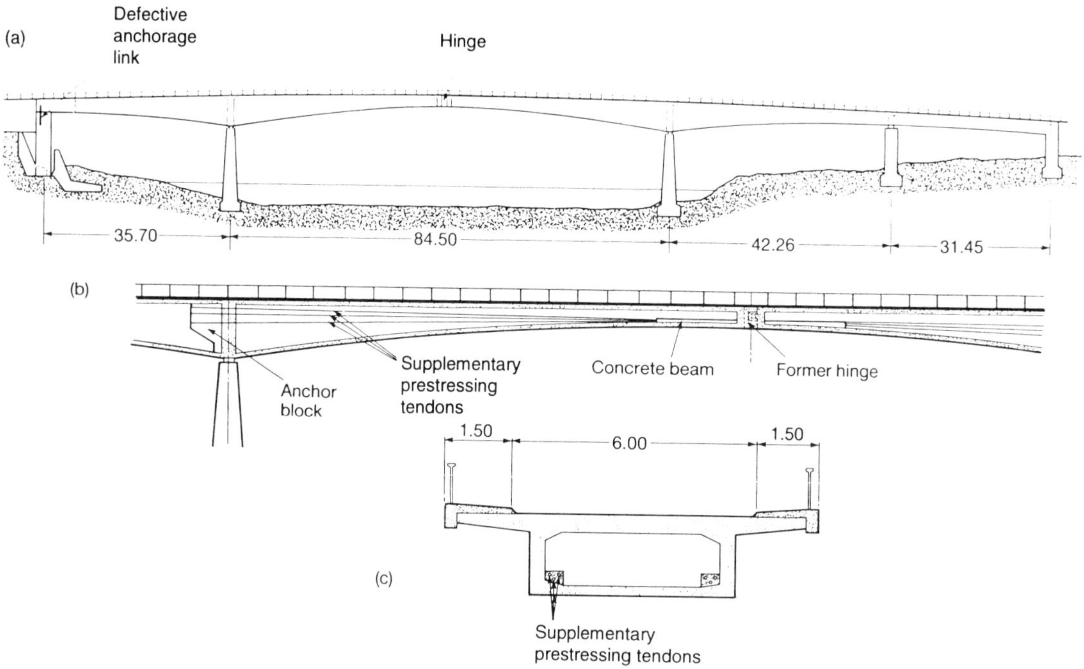

Fig. 2.4. Mosel Bridge:
(a) longitudinal section;
(b) supplementary prestress;
(c) cross-section

(3) Main Bridge

Built in 1878, the five steel framed arches were as shown in Figure 2.5, with eleven longitudinal bearers for the deck.

During the war, the middle part was strengthened by the addition of heavy reinforced concrete box beams underneath, retaining the steelwork under the roadway. Countless leakages in the concrete deck resulted in severe damage.

Rehabilitation carried out in 1987–88 included:

- reorganizing the pier area to reduce the transverse gaps from two to one,
- removing the old deck construction and providing a wider new in-situ deck slab for three lanes instead of two lanes.

This solution gave increased load carrying capacity and long-term improvement in durability and appearance.

2.2.15 Six bridges in Czechoslavakia—Dahinter (1990)

(1) Bridge over Sázava River near Kácov

The three-span continuous reinforced concrete girder bridge was completed in 1914. Water leakage and leaching combined with long-term carbonation led to badly corroded reinforcement and severe cracking, scaling, spalling and delamination of the concrete. All deteriorated concrete was removed, and reinforcement cleaned and replaced where required. Cracks were injected with a special epoxy solution, missing concrete was made good with epoxy mortar and

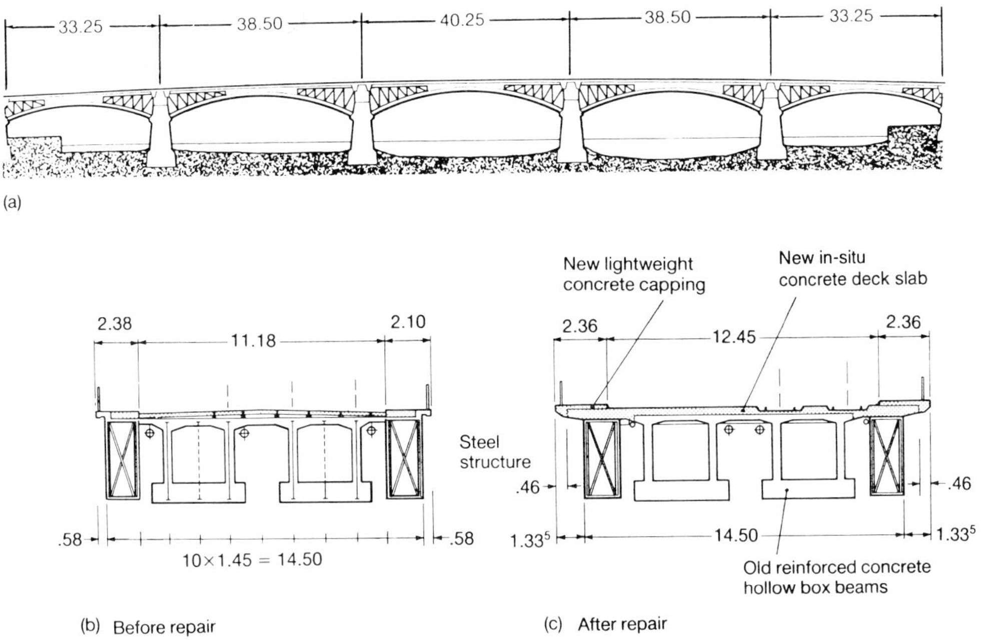

Fig. 2.5. Main Bridge: (a) longitudinal section; (b) cross-section before works; (c) cross-section after works

the deck strengthened with a 100 mm epoxy mortar overlay reinforced with welded mesh.

(2) Bridge over Ohře River in Karlovy Vary

The three-hinged reinforced concrete open spandrel arch 82 m span was completed in 1934. Later inspection recorded:

- ☐ a 100 mm deflection at the top hinge;
- ☐ water leakage, leaching, cracking and spalling, especially at the joints and the top hinge;
- ☐ local damage to arch ribs and deck supports.

Static and dynamic analysis found deflection due to creep and shrinkage and load carrying capacity adequate, subject to a speed restriction to 40 km/h for two years pending rehabilitation.

(3) Bridge over Berounka River in Beroun

The precast prestressed concrete I-beam deck was constructed in 1954 on piers and abutments which dated from 1864. After 30 years, the bearings, joints, waterproofing and surfacing needed attention but the superstructure concrete strength was at least 50 N/mm^2 and needed only local repair.

(4) Flyover crossing electrified railway in Pardubice

The 20 m span precast prestressed concrete structure was built in 1964. Special measures were needed to conduct a half-cell potential survey in the presence of strong stray currents. Cracks in the joints between box-beams impaired the transverse load distribution which was to be restored by reinforced concrete overslabbing.

(5) Bridge over Ohře River in Drahovice

The 66 m span cast-in-place prestressed concrete segmental box girder was completed in 1960. The development of a central 190 mm deflection, due to creep and shrinkage, led to ponding. Measurement of dynamic response in 1960, 1963 and 1987 found that the first natural frequency (2.40–2.33) and logarithmic damping decrement (0.055–0.036) were unchanged. The concrete was in very good condition and the only rehabilitation needed was to the waterproofing.

(6) Prestressed concrete frame bridge

The twin box girder bridge, completed in 1971, was cast in place on centring with 62 m centre span and an increased depth of haunch, after experience of the Ohře Bridge, (5) above. During construction, several cracks had appeared at midspan and later inspection noted that the cracks were closed with some signs of leakage. Otherwise the bridge was in good condition.

2.2.16 Performance of an 80 year old reinforced concrete bridge in an extreme environment—MacCraith (1990)

The single arch footbridge spanning 52 m about 45 m above the sea was constructed in 1909 at Mizen Head, County Cork. The arch ribs were precast and built out from the abutments without centring. The site is very exposed and sea spray has been known to reach above the deck level during severe southern gales. In 1972 the entire bridge was coated with bitumen with two layers of glass-fibre membrane and a top coat of masonry paint. In 1980 concrete around corroded bars was repaired with SBR polymer-modified cement mortar and made good with masonry paint. In 1986 some rust stained areas were treated with microporous paint to reduce ingress of chlorides. Tests in October 1989 found the concrete to be dense and of high strength, cover varied from 21 to 46 mm, carbonation depths averaged less than 1 mm and chloride ion concentrations 1.5% by weight of cement with some areas much higher, 4.6% from one core.

2.2.17 Elevated roadway at airport—Vaysburd (1990)

Rehabilitation of the bridge on the elevated roadway at the Baltimore/Washington Airport included repair of spalls along the expansion joints, the joints themselves and repair of cracked concrete. Compatibility with the existing concrete was the most important factor in selecting materials for repair.

2.2.18 A25 Woodbridge—Cogswell and Herbert (1991) and Department of Transport (1993)

This 21 m span reinforced concrete arch bridge was designed and constructed by the Trussed Concrete Company in 1912. It was treated extensively with sprayed concrete in the early 1960s. In 1988 a thorough investigation included a principal inspection, extensive testing and assessment of load capacity. Numerous areas of spalled concrete were found throughout the deck, due mainly to inadequate cover. There was no evidence of chloride attack and the strengthening work included casting a new reinforced concrete slab on top of the existing deck as shown in Figure 2.6.

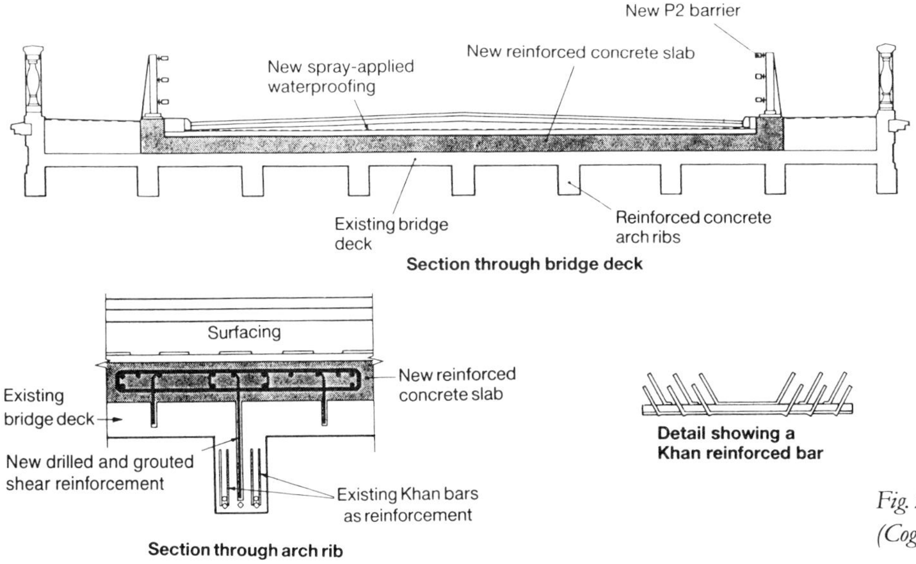

Fig. 2.6. Woodbridge, Guildford (Cogswell and Herbert, 1991)

2.3 Substructure

2.3.1 Pelham Bridge, Lincoln—Evans and Lee (1981)

The 350 m long structure built in 1957 comprises composite steel/concrete spans over the railway and in-situ concrete approach spans with 23 rows of columns. The deck was not waterproofed. Inspection in 1978 recorded deterioration of columns and deck due to leakage of the road joints.

A full corrosion survey included equipotential mapping, the chloride concentrations at the reinforcement and the chloride penetration gradient. Twelve columns were replaced with precast units up to 4.6 m long and 8 t in weight. After removal of faulty concrete and cleaning the reinforcement, 16 columns were refaced with superplasticized concrete pumped into the forms under pressure. The soffit repairs were similar.

Fine shrinkage cracks in the repair concrete were treated with silane in 1988 but monitoring had revealed no other problems up to 1991.

2.3.2 Cuyahoga River Bridge, Ohio—Freeh (1984)

Opened to traffic in 1955, the twin two-lane Ohio Turnpike bridges over the Cuyahoga River valley reach a height of 53 m above the valley. Each bridge has 4 × 30 m spans and 9 × 76 m spans with 12 reinforced concrete piers. As the use of deicing salts increased during the 1960s, so did the deterioration of the concrete portions of the bridges. In the mid 1970s efforts were made to divert the drainage and to patch the piers but the deterioration continued. By 1980 about 40% of the surface area of the piers was spalled or nearly spalled and thorough rehabilitation by shotcreting was proposed.

2.3.3 Coventry Inner Ring Road—Barfoot (1985)

A 1098 m long viaduct constructed in 1970 comprised precast prestressed box beams, jointed longitudinally with in-situ concrete and transversely post-tensioned, supported on reinforced concrete trestles. Road salt percolating through the expansion joints onto the tops of three crossheads caused spalling and cracking which was severe in places but localized and not threatening structural integrity in 1985 when repairs were put in hand. Up to 150 mm of contaminated concrete was removed from the sides and soffit edges of the crossheads, tackling the areas under alternate beams and making good with cementitious mortar containing a silicate additive. The infill areas were then repaired as a second stage. Leaving the bearings undisturbed kept down the cost of the operation but retained contaminated concrete, which was, however, contained under pressure. The cover was increased in certain areas using sprayed concrete reinforced with a polymer grid and the unrepaired concrete was sealed with a silicate solution to reduce its porosity.

2.3.4 Camsley Lane Viaduct, Cheshire—Ambrose (1985) and Thompson (1989)

The six reinforced concrete spans were constructed in 1963 to carry the M6 over the A56 and a railway. After 20 years the reinforced concrete substructure showed widespread staining, cracking, delamination and spalling as a result of chloride deicing salt leaking through the expansion joints.

Asphaltic plug joints were installed and the waterproofing and drainage of the deck were improved before tackling the repairs to the concrete. Carbonation depths were only 3 to 5 mm but chloride concentrations up to 5% by weight of cement were found near the surface and between 1 and 2% at depths between 100 and 150 mm. During the winter of 1984–85, with appropriate jacking support, concrete was removed to at least 20 mm behind the bars to expose at least 50 mm of uncorroded bar. After grit blasting, replacing bars of less than 90% nominal size, coating steel with an inhibiting primer and saturating the concrete, repair material was placed. A flowable concrete was cast against formwork in lifts up to 1.5 m and a repair mortar was applied in layers for areas less than 0.1 m^2.

As chloride contamination was still present, the effectiveness of the repairs were monitored by TRRL. Thompson (1989) found from surveys at intervals that after three years new corrosion sites had developed and that corrosion had continued in some repaired areas. He concluded that half-cell potential mapping alone was inadequate but, in conjunction with chloride concentrations, would lead to criteria with less risk of corrosion, albeit with more radical repairs and possibly more demanding temporary support.

2.3.5 British Rail Overbridge—Dixon (1988)

The bridge was constructed in 1966, comprising four prestressed spans supported by in-situ crossheads and columns. In 1985 cracking, staining and spalling was noted in 2 out of 3 crossheads and 5 out of 21 columns. In the course of conventional repairs with cementitious mortar with a track possession and scaffolding, deep cores and half-cell potentials were taken. Chloride contamination went through the full 1 m thickness of the crossbeam and half-cell potentials ranged from -450 to -550 mV. Exposure of the reinforcing steel revealed severe pitting and almost complete loss of section in places. The repairs were completed and a surface coating applied in order to slow down the deterioration, pending a permanent solution.

2.3.6 Midland Links Viaducts—Department of Transport (1985, 1988 and 1993), Boam and Unwin (1990), *Construction News* (1991) and *New Civil Engineer* (1992)

Background

The Midland Links interchange between motorways M5 and M6 in the West Midlands conurbation was completed in 1972. Some 21 kms of viaducts comprise simply supported composite steel beams and concrete decks carried on reinforced concrete crossbeams and circular columns. Spans vary from 15 to 45 m at an average height of 9 to 12 m above a network of roads, railways, canals and industrial premises. Maintenance problems have arisen with many of the 1200 crossbeams, 18 000 bearings, 3600 columns and 1800 expansion joints.

In the early 1980s deterioration was noted in the Principal Inspections and became a subject of concern. The Department of Transport convened a Working Party to:

- ☐ direct investigations,
- ☐ consider and agree actions to be taken and
- ☐ prepare a report and recommend future maintenance.

The Working Party published an Interim Report (1985) and a comprehensive updated Report (1988).

Early problems

- ☐ Bearing plinths with voids or damaged by water leakage.
- ☐ Buried expansion joints failed after about four years and nosing replacements could not be kept watertight.

☐ Deterioration of concrete due to leakage of salt-laden water.

In 1983, ten crossbeams were examined closely and found to have high chloride concentrations and associated reinforcement corrosion. An extended programme followed to investigate 44 crossbeams, a number of columns, foundations and parapet units; deck slabs were inspected in conjunction with resurfacing work. The investigations consisted of detailed visual surveys, testing for chloride concentrations, half-cell potentials and carbonation and petrographic examination of the aggregates.

Extensive chloride contamination was confirmed on the tops and sides of crossbeams, in shear walls and at the edges of deck slabs, all clearly associated with leakage from expansion joints. Main areas of deck away from joints were well protected by the waterproofing. There was extensive corrosion of reinforcement in crossbeams and shear walls, in some structures exacerbated by lack of concrete cover. Most of the corrosion was general with expansive corrosion products but some local pitting corrosion was also found. Carbonation was not significant. Concrete strength met structural requirements. Although aggregates generally were susceptible to alkali-silica reaction, one or both of the other joint causes (see 1.3.2) were insufficient for the reaction to be initiated.

Initial remedial work

Remedial work to bearing plinths was completed in a series of contracts from 1974 to 1984. Bearings were supported by jacks while voided or damaged concrete was removed either by mechanical tools or by water jetting. As salt-laden water was recognized as a major cause of the concrete damage, it was decided, after a successful trial in 1984, to use urea for deicing. It was also decided to replace the majority of the joints with asphaltic plug bridging joints across the full width of the structures. However, this work had to be phased with resurfacing over a number of years and meanwhile gutters were installed under the joints to collect and remove leaking water and, after joint replacement, to serve as a second defence. 92% of the joints had gutters and positive drainage by 1985.

Major maintenance programme

The deterioration of the crossbeams and columns was the most serious problem as it directly threatened the structural integrity of the viaducts. This, combined with the need to keep the traffic flowing and the scale of the rehabilitation work, called for the careful planning of priorities and the development of a long-term maintenance strategy.

Following the initial actions, a programme of monitoring the condition of more than 100 crossbeams was begun in 1985 and key elements were assessed to determine the effect of deterioration on their structural integrity. Trials were set up to test methods of repair, including concrete removal and replacement with flowable concrete. It was found that removal of concrete by high pressure water jetting has the advantage of not shattering aggregate. Also, it leaves a sound concrete surface and thoroughly cleaned reinforcement, prepared for casting the replacement concrete. However, the supply of large quantities of water to the equipment can present a problem and it may propel debris with great force. Therefore, for safety reasons, pneumatic hammers were allowed, as an alternative to water jetting, on spans over areas where the public had access.

Uncertainties about the most cost-effective option for repairing the viaduct crossbeams led to the development of a programme of repair trials. These included the following.

- *Concrete repair trials.* The removal of chloride contaminated concrete to differing chloride thresholds and replacement with flowable repair concrete.
- *Cathodic protection trials.* The application of a surface anode to chloride contaminated and corroding crossbeams through which a low voltage direct current is passed. In turn, this causes a current to flow through the concrete onto the reinforcement and thereby suppresses the current associated with the corrosion process.
- *Crossbeam replacement.* The demolition of existing crossbeams and replacement with beams of similar appearance designed to current standards.
- *Coatings.* Embedding monitoring equipment in six crossbeams to gather data prior to the application of surface treatments.

Doing nothing is also considered an appropriate option in the short term as it defers major expenditure. However, continued monitoring is required to ensure that structural integrity is not impaired.

Surface treatment is appropriate to beams in reasonably good condition whereas crossbeam replacement is only necessary for the very worst cases.

The need to undertake the work without affecting the traffic on the M5 and M6 motorways necessitated the provision of temporary steelwork supports for most concrete repair options and beam replacement. A typical sliproad crossbeam support for repair is shown in Figure 2.7.

Boam and Unwin (1990) described the maintenance strategy which was developed for the Midland Links structures. It was evident from the investigation work that, with some 1200 crossbeams in differing states of deterioration, it was essential to ascertain a cost-effective means of arresting the corrosion process and carrying out adequate long-term repairs. The paper illustrates the development of the strategy and how this was brought forward into a programme of repair works.

Construction News (1991) summarized the work at that time.

A crossbeam replacement reported in *New Civil Engineer* (1992) is shown in Figure 2.8.

This work was delayed while the explanation was discovered for an unexpectedly large thermal movement of the deck after the removal of the restraint of the shear walls. The excessive movement had resulted from the cumulative effect of continuity created by high frictional forces at the bearings of adjacent spans. After sliding surfaces were greased, thermal movements were within expected limits.

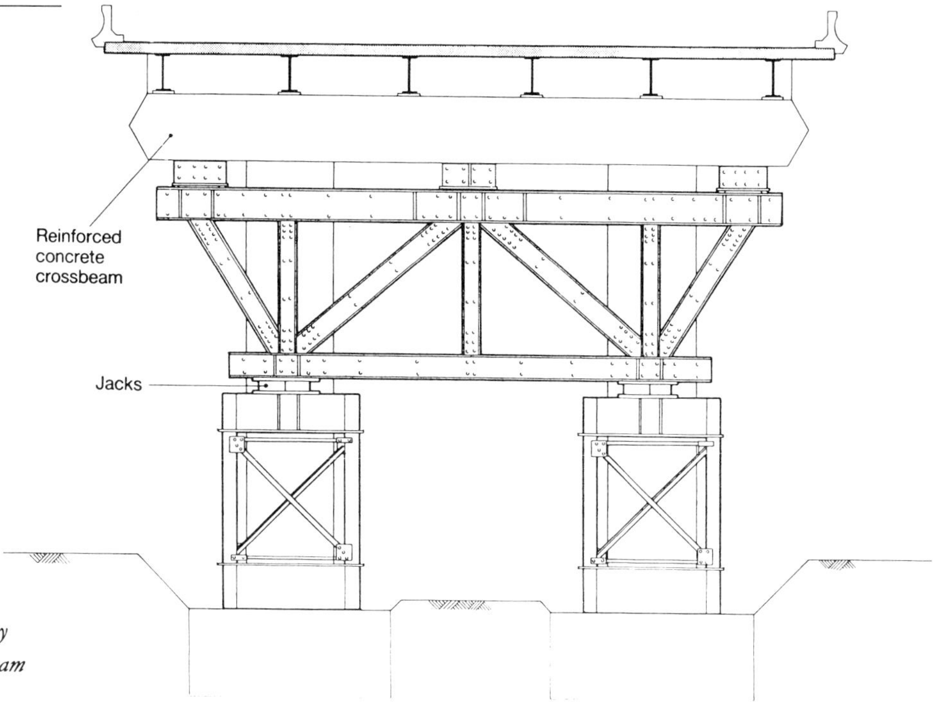

Fig. 2.7. Typical temporary support system for crossbeam (DoT, 1988)

2.3.7 Weaver Viaduct, Cheshire—Construction Repair International (1989)

The 270 m long prestressed concrete viaduct on six reinforced concrete cross-heads and piers was constructed in 1971 to carry the M56 over the River Weaver and the Weaver Navigation Canal. By 1988 salt-laden water seeping through the expansion joints had saturated the tops of the crossheads and caused corrosion of the reinforcement and spalling.

Rehabilitation entailed replacing contaminated concrete with a proprietary high-flow concrete, renewing 166 bearings and sealing the repaired crosshead with silane. At each crosshead in turn the section of motorway was supported on eight hydraulic jacks each of which carried 400 t, the traffic being uninterrupted except during the actual jacking process.

2.3.8 Repairs to Rakewood Bridge—Goldsby (1990)

In conjunction with extensive strengthening works to the 1969 structure required in the upgrading of the M62, repairs were necessary to the edge strips of all crosshead soffits. Access was by drop scaffolds hung from the crossheads but linked to a general access scaffold. Existing concrete was removed, using hand held pneumatic tools, to 25 mm behind the reinforcement, additional 25 mm bars were fixed in each area and the section recast with a high strength pourable grade repair mortar which was introduced through letter-box open-

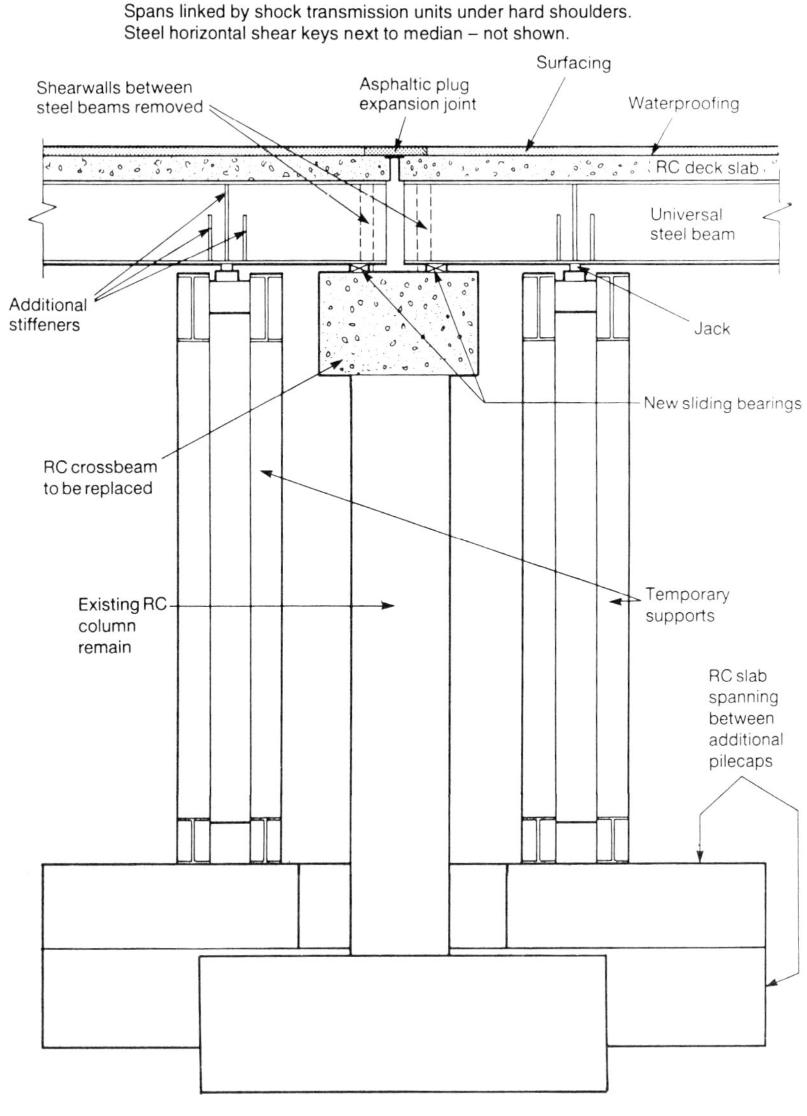

Fig. 2.8. Crossbeam replacement (New Civil Engineer, 1992, and DoT, 1993)

ings in the timber formwork. After at least 14 days curing, each repair was impregnated with monomeric alkyl alkoxy silane.

2.3.9 Crosshead repairs to M4 elevated section—*New Civil Engineer* (1991) and Department of Transport (1993)

The 2250 m long viaduct was constructed in 1965 and carries the elevated M4 motorway on the alignment of the busy A4 trunk road which is at ground level. The 17 m spans are simply supported on either side of reinforced concrete crossheads with expansion joints buried under the carriageway surfacing. The joints were not watertight and, despite remedial measures to reseal them, leakage of salt-laden water had damaged the 104 reinforced concrete crossheads. Most were affected by staining, cracking and spalling to differing extents and some previous patch repairs had started to show signs of distress. An investigation is in progress to determine the extent of rehabilitation work on the crossheads which overhang two of the three lanes of each carriageway of the A4.

2.3.10 Wardley Hall Viaduct—*New Civil Engineer* (1991) and Thoday, Jones and Simpson (1993)

The 292 m long twin eight span structure, built in 1970, provides the important link between the M61 and M62. The ends of the 16 simply supported plate girders were pinned together, provision being made for expansion at one abutment only. The rotation joints and a longitudinal joint leaked from the start and an inspection in 1985 recorded extensive spalling in the pier hammerheads. About 30% of the surface area was affected and tests found chloride concentrations up to 3.5% by weight of cement and up to 50% loss of section of shear steel.

Each of the 14 hammerheaded piers had to be partially or fully replaced without affecting the motorway traffic. To achieve this, a temporary support system was installed at each pier hammerhead in turn. Sixteen separate steel tube props, individually jacked and computer controlled to ensure that none of the 16 deck girders was lifted more than 2 mm, carried the loading while the hammerheads and, at two locations, piers were demolished and reconstructed. Most of the demolition, which amounted to 50% of the total substructure, was done with jack hammers but it was economic to employ an ultra-high pressure water jetting system operating at 248 N/mm^2 and with estimated nozzle speeds approaching 4800 km/h.

Thoday, Jones and Simpson (1993) reported the investigation, repair and strengthening in detail.

2.3.11 Öland Bridge, Sweden—PIARC (1991) p 321, Von Olnhausen

Built 1967–72, the five span high-level section is 910 m long and rises to 36 m at the navigation channel. Low-level approaches consist of 23 spans of 35 m to the mainland and 124 spans of 35 m to the island of Öland. The overall length is 6072 m and the width is 13 m, comprising two 3.5 m traffic lanes and two 3.0 m emergency lanes. High quality 40 N/mm^2 concrete with air entrainment, the deck was waterproofed with mastic asphalt before laying the hot-rolled asphalt surfacing and exposed concrete surfaces were protected with silane. The whole length of superstructure and the substructure of the high level section have remained in good condition. On the Öland approach, however, less than ten years after construction large cracks developed at the tops of the cross beams (hammerheads) and the reinforcement of the piers suffered severe corrosion. Although the salt content in the Baltic Sea is only 0.3 to 0.4% at the bridge, salt concentrations up to 5% were found in the splash zones. Half-cell potentials showed that reinforcement was corroding as much as 5 to 6 m below water level under the porous but still intact concrete cover. Above water the cover was spalling. After careful consideration of different methods of rehabilitation, some of which interfered with traffic, it was decided to encase the whole of 105 existing piers and hammerheads in 400–500 mm thick concrete. Work proceeded in caissons with no restriction of traffic.

2.4 General deterioration

2.4.1 Tay Road Bridge, Scotland—*New Civil Engineer* (1984, 1988 and 1989), Maxwell (1985), Watters (1991), Crerer (1993) and Shields (1994)

Constructed in 1966, the twin steel box girder superstructure has a composite concrete deck and the 42 spans are supported by 41 pairs of parabolic section solid concrete columns rising to a height of 30 m. In 1983 cracking and spalling was reported in the lower sections of several mid-river columns but access was not adequate for a full inspection at that time. However, the integrity of the lower 4 m of all columns was tested by tapping with a hammer and all had a hollow sound. Core tests indicated that concrete strengths were up to 50 N/mm^2 although cover to reinforcement was as little as 20 mm in places. Concrete deterioration was attributed to the seawater and deicing salt leaking through the carriageway joints. Close to the bridge site, coating tests were set up on 15 concrete manhole rings of the same circumference as the columns, some combined with cathodic protection, and effort was concentrated meanwhile on the provision of an effective access gantry to enable the permanent repairs to the superstructure to be undertaken.

Maxwell (1985) described the investigation and the following proposed remedial works:

- replacement of inspection/maintenance gantries;
- deck concrete repairs, joint replacement and deck waterproofing;
- external/internal shot blasting and painting;
- repairs/remedial works to concrete substructure.

New Civil Engineer (1988) reported the proposed remedial works to include a new concrete skin for probably all of the 82 columns in addition to deck waterproofing, replacing all expansion joints and resurfacing. Although the access gantry was operational the work had to await the next weather window in the spring of 1989.

Reporting the start of work, *New Civil Engineer* (1989) noted the decision to protect cathodically the bottom 6 m of all columns. 1.5 mm wire mesh was to be wrapped around the columns to form the anode and encased within a 50 mm overlay of high bond, high flow, non-shrink, low resistivity concrete mix. Four proprietary materials were to be tried and two columns were to be treated with no-flow sprayed mortar.

Development of the cathodic protection for the substructure was described in detail by Watters (1991) and the following situation report was provided by

Crerer (1993). The following works were programmed for completion by the end of the summer season 1993:

- ☐ provision of two new inspection/maintenance gantries and 4500 m of special section runway beam;
- ☐ replacement of all 92 bridge expansion joints and 34 joints on the elevated approach ramps;
- ☐ hand and sprayed mortar repairs to bossed and spalled concrete on all elevated approach ramps, including the application of a protective system to reinforcement;
- ☐ installation of cathodic protection to the bottom 6 m of 75% of river piers, using coated expanded titanium mesh encased in a 50 mm protective jacket of flowable concrete poured into formwork in 2 m lifts;
- ☐ patch repairs to tops of river piers;
- ☐ external and internal painting of 50% of the steel box girders and transverse diaphragms;
- ☐ removal, refurbishment and refixing 4500 m of aluminium vehicular parapet and copings;
- ☐ construction of the new Toll Plaza Canopy, installation of toll collection equipment and demolition of the old toll booths and elevated observation platform.

Shields (1994) confirmed completion of the above, except some 10% of the cathodic protection of the river piers and further ongoing concrete repairs in the area of the central walkway.

2.4.2 Pag Bridge, former Yugoslavia—Sram (1985 and 1987)

This concrete structure, connecting the island of Pag with the mainland, is situated in a very aggressive marine environment. Deterioration could be attributed to faults and negligence in design or construction or to imperfect technology. Sram (1985) analysed the possible causes and pointed to the fact that damage is rarely due to a single cause. Sram (1987) described the corrosion of reinforcement in the deck girders and the two solutions to the problem. Repairing the existing structure had certain advantages but might have led to further unjustified investment if it had failed. He therefore proposed a method by which it was possible to replace the deck gradually without stopping the traffic.

2.4.3 Tees Viaduct—*New Civil Engineer* (1988 and 1991) and Johnson (1993)

The 2 km long high-level viaduct over water and railways, was constructed in 1975, comprising 5 steel plate girder spans and 63 simply supported approach spans, comprising 941 mm deep universal beams with a composite reinforced concrete deck, all supported on reinforced concrete crossheads and piers.

A comprehensive investigation undertaken in 1987 revealed typical carbonation depths of 2 to 3 mm. However, the effects of deicing salt, heavy traffic and the industrial pollution had resulted in deterioration and the need to replace 565 bearings, nearly all of the 120 expansion joints and to repair sections of the deck. After experience of unsuccessful trials of replacement joints in the 1970s it was decided to tie the decks at alternate supports and to install Maurer expansion joints at the others. Patch repairs were carried out on the deck which was then waterproofed with a sprayed membrane and resurfaced.

New Civil Engineer (1988) described the concrete repairs and enclosure of the superstructure, using a specially developed GRP panel system. This protected the steel plate girders and bracing and the deck soffit from the aggressive environment and provided a ready means of access to the structure for the repairs and for future inspection and maintenance.

New Civil Engineer (1991) reported that half of the piers had chloride concentrations in excess of 0.4% by weight of cement and ten of them had to be replaced. A 10 t section from one of the crossheads had been sent for trial desalination, one of the remedial options under consideration.

2.4.4 Harry W. Nice Memorial Bridge—Lutz (1990)

After 43 years in service, the two-lane bridge carrying Highway 301 over the Potomac River needed rehabilitation. Maryland Transportation Authority also proposed to strengthen and widen the bridge, for which preliminary consideration suggested that there would be 45 months of traffic disruption. During design, innovative alternative concepts were proposed and accepted which reduced the construction time and cost and the loss of tolls significantly.

2.4.5 Three bridges in Saudi Arabia—Al-Babtain and Abbas (1991)

Salt contaminated aggregates were quite widely used in Saudi Arabia and there was a tendency to use high water/cement ratio to compensate for loss of workability at high temperatures. These factors, combined in some cases with poor quality and inadequate thickness of cover, resulted in loss of durability, mainly due to corrosion of the reinforcing steel. The paper was concerned with three bridges which were showing signs of concrete distress.

(1) Abu-Maan Bridge

The in-situ voided deck had deteriorated due to the effects of chloride concentrations up to about 4% by weight of cement. More than 80% of the deck area was delaminated and removal of the cover revealed complete disintegration of the top steel. After providing temporary support, the top 200 mm of the deck slab was removed and new steel and concrete placed using a bonding agent.

(2) Al-Nabya Bridge

There were signs of distress in the deck slab but it was found that chloride concentrations were less than 1%, although well above the usually accepted corrosion threshold of 0.2% by weight of cement. A half-cell potential survey found span 2 to have 27% of its area more negative than -350 mV and a reasonably well correlated 32% delaminated. Of this area, 13% was more negative than -500 mV, suggesting the risk of pitting. Span 4 had 43% of measured potentials more negative than -500 mV. Breaking out confirmed an average loss of section of 30% but the worst locations showed corrosion down to 10% of section or total loss. Additional steel was added where required and the repaired deck slab was then prepared for cathodic protection.

(3) Aramco Bridge

The reinforced concrete pier and pilecap was in a similar state to the Al-Nabya Bridge deck and treated similarly as a cathodic protection trial.

2.4.6 Polcevera Viaduct, Genova—PIARC (1991) p 93, Malisardi and Rigacci; Camomilla, Donferri and Grenari (1993)

Built in 1967, the structure has needed rehabilitation of piers, prestressed T-beam decks, balanced cantilever box structures, pylons and stays. Side cantilevers were reconstructed and bearings and joints replaced. Movable access equipment enabled work to proceed without obstructing traffic.

Camomilla, Donferri and Grenari (1993) reported on dynamic and reflectometric measurements on the stays.

2.4.7 Teiro Viaduct, Genova–Savona Autostrada—PIARC (1991) p 105, Malisardi and Rigacci

Built in 1960, the structure has needed the rehabilitation of ten reinforced concrete spans, comprising box-cantilevers with suspended T-beam decks on Gerber half-joints, and reinforced concrete piers. One girder at a time was supported on a temporary steel girder by Dywidag rods at the Gerber joint. The top 500 mm of the cantilevers were demolished and recast and bearings replaced. Two lanes were available for traffic throughout the works except for brief reductions to one lane. A travelling gantry with a platform giving access to the soffit had a rotating feature enabling it to pass the piers.

2.4.8 Aglio Viaduct, Bologna–Florence Autostrada—PIARC (1991) p 115, Romagnolo

Built in 1960, the reinforced concrete spandrel piers to the major twin arch and the piers to the 18 approach spans had deteriorated so badly that it was decided to encase them in 120 or 150 mm thick rheoplastic shrinkage-compensated concrete. Local repairs were required to the arch and the decks. Bearings were upgraded to meet current seismic requirements. Provision of a 1.6 km access road and some considerable investment in concrete pumping and material handling equipment amounted to about a third of the estimated cost to road users and the community if the motorway had been obstructed.

2.5 Post-tensioning tendons

2.5.1 Braidley Road Bridge, Bournemouth—*New Civil Engineer* (1978 and 1980)

The concrete box girder structure was designed in 1966 and completed in 1970 with unbonded post-tensioning tendons. The 120 external tendons comprised 19-wire 28 mm diameter strands beside, not within, the webs. The cables were protected with an aluminium coat inside a PVC sheath but fractures started to occur due to stress corrosion pitting and it was decided progressively to detension, remove and replace all of the tendons and encase them in concrete. The techniques were described in *New Civil Engineer* (1978) and, in a review of the merits of unbonded tendons (1980), it was reported that the replacement of tendons at Braidley Road was completed by October 1979.

2.5.2 A3/A31 Flyover, Guildford—*New Civil Engineer* (1980 and 1994)

Built in 1975, the 20 m and 50 m spans are of continuous segmental construction having a single cell box section spine beam with cantilevers. It is longitudinal post-tensioned with unbonded strands inside the box between anchorages at the abutments and deflected on saddles at the pier diaphragm. Three years after construction, cracking was found on the inside of the 20 m span anchorage diaphragm which gave cause for concern and some less extensive cracking at the other end diaphragm. The bridge was closed pending repairs. Fortunately it was lightly trafficked and the temporary arrangements caused no great problem.

On investigation the cracking was found to be due to a deficiency of reinforcement. The first stage of the remedial work was to apply vertical compression by means of unbonded bars stressed against temporary steel yokes. The second stage, the permanent repair, required casting additional concrete at both anchorages and the intermediate diaphragm and providing orthogonal post-tensioning which restored the structure to the serviceability originally intended.

New Civil Engineer (1994) reported the temporary closure of the bridge on account of a number of broken wires due to corrosion at anchorages. The 50 m span was propped pending the consideration of replacement options and the permanent solution of the problem.

2.5.3 Repair of Fives Road Bridge—Borderie (1981)

The three span continuous post-tensioned concrete structure was built in 1956. Following an inspection in 1966 which noted cracking, a full investigation was put in hand. This included levelling, acoustic and gamma-radiographic examination and making some openings for direct inspection of the tendons. More than half of the ducts inspected were incorrectly grouted, wires were corroded and some were fractured. Short-term repair consisted of injecting the cracks, reinforcing the webs and splinting the joints. The long-term repair, completed in 1977, involved the provision of additional longitudinal post-tensioning.

2.5.4 Repair of Belgian Road Bridge—Kreps-Hendrikx and Callens (1981)

The continuous bridge was constructed in 1960 with a central span of 52 m and two side spans of 26 m. External prestressing tendons comprised wires anchored by the Magnel 'sandwich' system. Each pair of tendons consisted of 104 wires which were brought together to form a single bundle at midspan between the inverted T-beams, encased in cement mortar, their only protection from the environment. With no waterproofing on the bridge deck, the wires became so corroded that all of the cables had to be replaced in 1976. New BBRV tendons were protected by PVC sheaths and cement grout.

2.5.5 Seventeen Ontario bridges—Ryell, Manning and Owens (1984)

In a 1974 study, involving 15 structures, 150 mm diameter cores were taken through cracks in the deck slab where the ducts were closest to the road surface. When the ducts were located, portions of sheath and exposed grout were removed to inspect the cable strands. No corrosion of the cables was observed and with one exception the ducts examined were full of hard, sound grout. The deck slabs had exposed concrete roadway surfaces with cover to the ducts from 76 to 230 mm.

More recently, cables removed from a pedestrian overpass constructed in 1962 were severely corroded. There was a significant loss of cross-sectional area of some wires where grout did not fill the duct or was of very poor quality.

A bridge built in 1958 was demolished in 1981 for a highway realignment. Cables removed from the cast-in-place hollow box deck slab had only isolated examples of slight surface corrosion on the wires. The ducts were fully grouted with dense, high quality Portland cement grout.

2.5.6 Gardiner Expressway, Toronto—Tork (1985)

The 18 to 30 m simply supported prestressed concrete box beams were either pre-tensioned or post-tensioned with strands. Full investigation after 16 years' service found rusted reinforcing steel, rusted prestressing strands and delaminated or spalled concrete. One beam suffered 60% strand loss. The causes of deterioration are discussed and repair methods proposed.

2.5.7 Taff Fawr Bridge redecking—*New Civil Engineer* (1985)

Constructed in 1964, the four webs for the continuous 58 m river span and two 39 m side spans were precast I-section segments cantilevered from the piers and stressed back in turn. In situ 200 mm infill slabs were cast to form top and bottom flanges of the box section. The principal inspection in 1983 noted spalling from lower flanges and, in the course of repairs, serious corrosion of the prestressing tendons was discovered. The ensuing intensive investigation attributed the trouble to salt-laden water percolating through joints in the waterproofing. Alternative schemes were drawn up for partial or complete reconstruction of the deck while maintaining the traffic on the busy A465, Heads of the Valleys Road. The successful bid was for the complete demolition of the existing deck and the construction, in balanced cantilever, of a new single box in-situ structure. The work commenced in 1985 with the launch of a 305 t temporary bridge, 155 m long in four spans.

2.5.8 Pancevo Bridge, former Yugoslavia—Djurdjevic and Pakvor (1985 and 1989)

The 82.5 m continuous three span prestressed box-girder bridge was constructed in 1964, one of sixteen interconnected access structures forming part of the Danube crossing in Belgrade.

In 1976 cracks in the vicinity of the supports and at midspan were injected with epoxy resin. In 1984 corrosion of the reinforcement and spalling in the bottom slab of the box was such that the slab required replacement and the prestressing tendons had to be replaced. The deterioration and repairs were described, including a special structure forming part of the deck expansion joint (1985).

A further detailed inspection and analysis of the structural complex revealed cracking as shown in Figure 2.9(a).

Fig. 2.9. Pancevo Bridge, Belgrade: (a) location of cracking; (b) supplementary prestressing (Pakvor and Djurdjevic, IABSE, 1989)

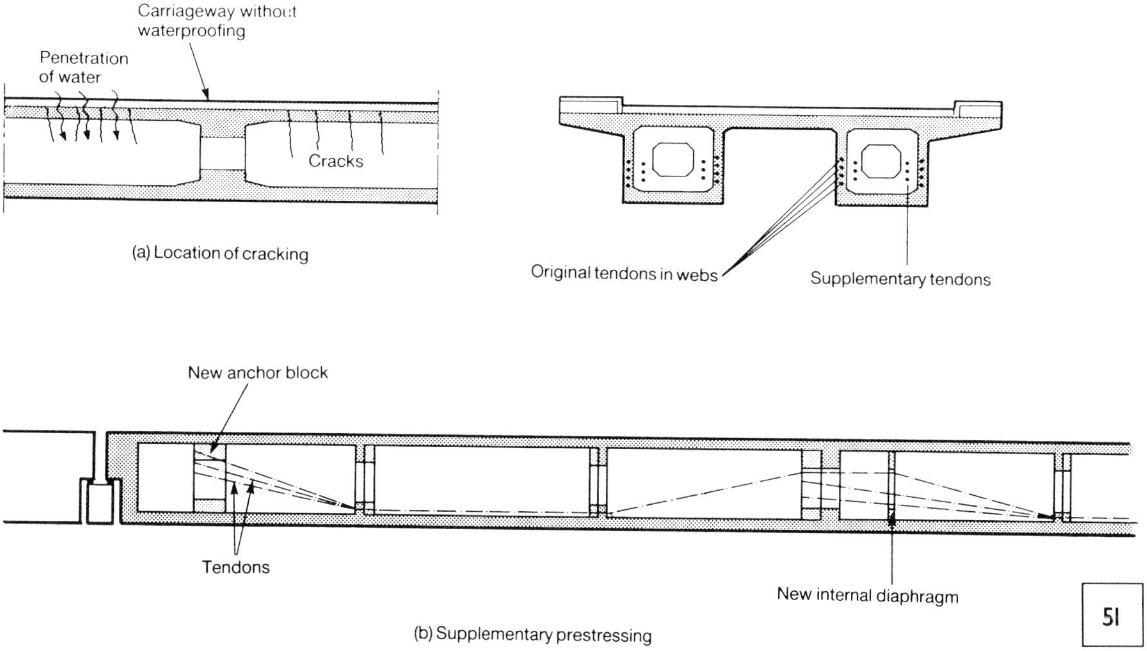

The rehabilitation described (1989) included the carefully controlled pressure injection of cracks wider than 1 mm with low viscosity epoxy resin and supplementary prestressing which ranged from 60 to 90% of the originally designed prestressing force according to the capacity of the section. A typical arrangement of the supplementary prestressing is shown in Figure 2.9(b). The opportunity was taken, at the additional anchorages and deflectors, to provide for a further 25% of additional tendons, should these be required at some time in the future. Traffic was completely stopped while the crack injection work was in progress but, otherwise, traffic was maintained on one side of the bridge. Two of the structures needed temporary supports.

2.5.9 Road bridge over D-T-D Canal, former Yugoslavia—Gogic and Mozina (1985)

The stability of the box-girder bridge carrying the Novi Sad–Temerin Road was affected by severe corrosion of prestressing cables. The repairs were described.

2.5.10 Wentbridge Viaduct—*New Civil Engineer* (1986) and Department of Transport (1993)

Completed in 1961, the three span structure carries Trunk Road A1 over the River Went in West Yorkshire. The 6-cell box-section superstructure was cast in situ and post-tensioned with 12 tendons, each comprising 17×19-wire strands, external to the webs, one tendon to each cell wall. It lies at the bottom of a sag curve in the road alignment and in 1985 a principal inspection identified chloride damage due to water entering access manholes and ponding inside the cells. Pending repairs, a weight restriction of 120 t was applied; when further investigations revealed tendon corrosion, this was reduced to 75 t, compared with 300 t original design capacity.

As a first stage in the rehabilitation, new 700 mm diameter access holes were cut through the webs and a walkway provided under the eastern cantilever between abutments. The manholes in the deck slab were then sealed with concrete and the deck waterproofed and resurfaced. The condition of the tendons was checked by selectively breaking out their protective concrete encasement and examining them, especially adjacent to disused access manholes, anchorages and deflectors. A full corrosion survey revealed that the damage to the tendons was not as extensive as had been thought initially and, after a full structural assessment, the weight restriction was lifted. Further rehabilitation works continued.

2.5.11 M24 bridge over D-T-D Canal, former Yugoslavia—Milanov (1988)

The 220 m long structure near Melenac suffered deterioration of concrete and corrosion and fracture of prestressing tendons due to the penetration of water and salt. Details were given of the different stages of the repairs including temporary and final tensioning, anchorage and reinforcement of slabs and transverse girders.

2.5.12 Prestressed concrete pedestrian bridge—Jovovic (1989)

This bridge was constructed with less prestressing tendon section than was required for the design. Three options were considered:

(*a*) insert more tendons into the bridge section
(*b*) add tendons outside the bridge section
(*c*) glue steel plates to the soffit.

Option (*b*) was selected.

2.5.13 Prestressed concrete girders after 20 years—Leon, French and Olson (1989) and Olson, French and Leon (1990)

Four AASHTO Type iii prestressed bridge girders, cast in 1967, and placed in an interstate overpass, became available for testing. The objectives were:

(*a*) to compare actual prestress losses with code values,
(*b*) to compare material properties found by NDT with cores,
(*c*) to examine performance under various fatigue loadings,
(*d*) to determine sustainable impact damage before repair,
(*e*) to evaluate fatigue performance of strand splice and external post-tensioning repair schemes.

Prestress losses were approximately as predicted by US codes. There were few signs of any corrosion or durability problems.

2.5.14 Ynys-y-Gwas Bridge—Woodward and Williams (1989)

The single 18 m span segmental structure was built in 1953 and collapsed in 1981. In the subsequent investigation it was found that chloride contamination due to deicing salt at the joints between the segments had caused local corrosion and fracture of the tendons.

2.5.15 Blackburn Road Bridge—*New Civil Engineer* (1990) and Department of Transport (1993)

Constructed between 1966 and 1968 to carry the northbound entry slip road onto the M1 at J34, the five span structure comprised precast longitudinally post-tensioned beams with an in-situ concrete slab which also contained post-tensioned tendons and anchorages. In the course of a special inspection, half-cell potential readings indicated the probability of corrosion of reinforcing steel in the deck slab. Further investigation revealed corrosion in the vicinity of anchorages in the top slab and in 1990 an ungrouted, untensioned tendon was found. Following this, all 120 tendons in the deck slab above the piers were investigated; 41 were found to be ungrouted and 15 partially grouted. Three tendons, all above pier No. 3, were found to be untensioned, having completely corroded through near the dead end anchorages. Other tendons exhibited general surface corrosion and pitting corrosion.

In addition, some 30 tendons in the main beams (about 30% of the total number) were investigated and, of these, 11 were found to be partially grouted and corroded in places. The transverse post-tensioning was also investigated as far as was practicable. The bridge was then propped to allow it to remain in service while a feasibility study was undertaken. The deck was replaced in 1992.

2.5.16 Long term behaviour of materials in a prestressed concrete bridge—Isecke (1990)

Results are given from material probes during demolition of Schmargendorf Bridge on the Berlin Autobahn Ring. The bridge, constructed in 1960, showed numerous longitudinal and diagonal cracks in 1969 and required extensive rehabilitation. During a main inspection in 1977–78 additional failures were found and it was decided to demolish and rebuild. Salt-laden water had flowed through an expansion joint in the deck and over the unprotected prestressing anchorage which was destroyed. Ingress of chloride caused fracture of prestressing steel in two locations. Depths of carbonation were between 1 and 3 mm.

2.5.17 Evaluation of old prestressed concrete bridge—Tassios and Aligizaki (1990)

A 30 year old 140 m long prestressed concrete bridge was inspected and evaluated. In-situ and laboratory tests were carried out to determine strength, carbonation depth, chloride content, static and dynamic loading characteristics. The main findings were that:

- the quality of concrete remained good;
- corrosion had destroyed some tendons;
- some bearings were compressed.

2.5.18 Two N9 bridges near Lausanne—Fuzier (1990)

Constructed between 1971 and 1973, two parallel bridges about 400 m long, each comprising three portals on twin column piers, carry the N9 motorway over the Lutrive Valley. Inspection recorded the loss of the initial camber and checks showed that the deflection was continuing to increase. In 1986 the behaviour of the structure was analysed and the remedy prescribed. Work to install 18 tendons throughout the whole length of the structures included drilling holes for the passage of tendons and bars, fixing deviators and anchorage blocks with Macalloy-Freyssinet bars and the completion of the stressing. This was done without interruption of traffic. The rehabilitated structure was monitored for six months and then the tendons were grouted with cement grout.

2.5.19 Ortaglia, Alteta and Ginepro Viaducts, A1 Milan–Rome—PIARC (1991) p 81, Paccagnella

Built in 1960, the structures needed rehabilitation of twin prestressed T-beam decks, augmenting both the reinforcing steel and prestressing steel. Deteriorated concrete was removed by hydrodemolition which resulted in good adhesion of the new concrete. The depth of carbonation was 20 to 30 mm. Transverse prestressing tendons were cement grouted and the longitudinal tendons were epoxy coated, greased and sheathed. The only interference with traffic was two days' reduction to one lane for each span.

2.5.20 Structure OA340, Ferney–Versoix, Switzerland—PIARC (1991) p 359, Canton Geneva

Built in 1962–64 to carry Highway N1, the 60° skew deck slab is 630 mm thick, with square span 15 m and the length of free edges 33 m. The deck was constructed monolithically with twin walled abutments. In 1962, when the deck soffit formwork was removed, a number of cracks appeared and the free edges deflected 30 mm. To remedy this, 47 post-tensioning cables were installed but these early disorders contributed to the premature ageing of the structure. By 1985 the slab had cracked further with rust staining, the post-tensioning cables were severely corroded and the free edge deflection had increased to 84 mm. The concrete, though cracked and carbonated deeper than the bars, was of good quality and the average strength from 47 cores was 85 N/mm^2. It was decided that the best rehabilitation technique was to reinforce the slab with glued steel plates. Traffic on Highway N1 was hardly affected and the single lane interventions for five months were acceptable on the minor road under the bridge.

2.5.21 Sixteen English bridges—Woodward and Loe (1981), Woodward (1981), *New Civil Engineer* (1992) and Darby (1993)

Woodward and Loe (1981) described the inspection of two post-tensioned bridges constructed in 1961. In 1970 longitudinal cracks following the tendon profile were noticed and a full inspection revealed cracks in 55 of the total of 144 beams. Holes drilled into the ducts (see 3.4.1) proved the presence of voids and water in the ducts. The pH of the water sampled was higher than 11.5 and the chloride concentrations were very low. Subsequently, an exploration with a borescope showed, at most locations, a film of cement paste and no significant corrosion on the surface of the wires.

The second structure showed no external evidence of trouble. The beams were post-tensioned longitudinally through steel lined ducts and transversely through diaphragms having unlined ducts. Careful investigation by hole drilling showed that 62% of the ducts had voids along their whole accessible length and two ducts contained no grout at all. Of the lined longitudinal ducts, 28% contained moist grout and 9% contained free water. The unlined ducts were drier; the condition of the wires in the two ungrouted ducts varied from nearly perfect to extensively corroded and several wires were found to have fractured at locations where water and air were present.

Woodward (1981) described an investigation of the grouting of twelve post-tensioned bridges built between 1958 and 1977 in Oxfordshire, Kent and Hereford and Worcester. Ten bridges were examined in situ and the other two had been demolished. Drilling into ducts, voids were found in ten of the bridges including the one most recently built. Some voids were found to be continuous along the accessible length of the duct and they tended to concentrate at high points. Local voids were found at anchorages of beams with otherwise fully grouted ducts. At six bridges, the voids were of sufficient size to expose the tendons partially, which were mostly covered with cement paste and with no evidence of serious corrosion. The one exception was a strand with slight pitting in a virtually empty duct. The presence of some grout in the duct provided an environment in which corrosion did not take place but this protection could be changed by the entry of air, water and chlorides. The results of this work were presented to the Standing Committee on Structural Safety which concluded that, in structures containing a large number of tendons: 'the risk of sufficient tendons failing by corrosion at any time to cause sudden collapse is considered small'.

New Civil Engineer (1992) reported the defective central span of the 30 year old Botley Flyover, carrying the A34 trunk road in Oxfordshire. Longitudinally and transversely post-tensioned trapezoidal beams with mortar infill were investigated after chloride contamination was discovered. This involved drilling into eight longitudinal ducts at twelve locations and exposing four end anchorages. All ducts contained negligible amounts of grout and three of them contained water. Concrete surrounding the end anchor plates was of poor quality and contaminated with chloride. Tendons were in excellent condition at all locations except for very localized pitting adjacent to the end plates. Only two wires had fractured but the anchorage collets were so seriously corroded that they had lost their separate identities. The span was propped until replacement in 1993.

Darby (1993) provided particulars of Folly New Bridge, Bladon, which was constructed in 1953. The single 13 m span comprised segmental hollow rectangular beams (with no in-situ top slab) prestressed by twelve 7 mm Freyssinet tendons with anchor cones. Longitudinal tendons were ducted in the first segment but for the rest of their length they were external to the beams and surrounded by cement rich sand/cement grout. The transverse tendons were grouted. Tests in 1988 found 2.75% chloride by weight of cement in the grout surrounding the wires. The wires were found to be pitted, severely so in honeycombed locations. Some of the short ducted sections were ungrouted with more than 50% of the wires fractured. Transverse ducts had been well grouted and the wires were in good condition. The bridge has been demolished.

2.6 Repair of bridges under traffic—PIARC (1991)

The PIARC Technical Committee on Bridges reported 33 case histories of many types of bridge work where interaction with traffic was of special interest.

Examples of concrete bridges are summarized below with page numbers and authors from the PIARC Report.

2.6.1 Autoroute A6 viaduct over River Bievre, South of Paris—PIARC (1991) p 9, Picard and Chatelain

Built in 1959, the bearings were replaced in 1986. To achieve this the prestressed concrete deck was lifted by ten jacks, on temporary piers, with automation to prevent the difference in level between adjacent girders exceeding 0.3 mm. The 28 deck lifting and lowering operations were carried out with the bridge closed to traffic for less than one hour, at night.

2.6.2 Expressway over River Seine at Corbeil—PIARC (1991) p 15, Faveau and Chatelain

The three span segmental twin box girder bridge was built in 1973. To provide additional prestressing, massive in-situ anchorage blocks were constructed, while traffic was carried on the other girder. Joints were grouted and cables tensioned during short bridge closures when the traffic intensity was low.

2.6.3 Fiumicello Viaduct, Autostrada A1 Milan–Rome—PIARC (1991) p 69, Paccagnella

The structure was built in 1960 and rehabilitated in 1985–86 in order to upgrade seismic protection. The 16 twin simply supported reinforced concrete T-beam decks, forming a 55° arc with 300 m radius, were made continuous and abutment restraint was increased to restrict seismic stresses on the piers. Work was completed without closure to traffic in either direction. There were lane restrictions but one lane was always kept clear for emergency traffic on the structure having work in progress.

2.6.4 San Milano Bridge, Bruncu Teula, Sardinia—PIARC (1991) p 137, Rusconi and Atzeri

Built in 1936, severe deterioration of this five span reinforced concrete bridge was found, especially at the Gerber half-joints of the central span. Carbonation was 50 to 60 mm deep and about half the cross-sectional area of the reinforcement had been lost by corrosion. Repair involved the removal of poor concrete, adding new reinforcement, placing new concrete cover and renewing the bearings. To avoid lifting from above and obstructing traffic, the suspended span was jacked against temporary steel girders underneath as shown in Figure 2.10.

2.6.5 Miniway, Hanshin Expressway, Osaka—PIARC (1991) p 173, Okuda

A temporary overpass, Miniway, was developed to maintain traffic on one lane for localized work to be carried out underneath or alongside. Miniway comprises a working module, three ramp-up modules and three ramp-down mod-

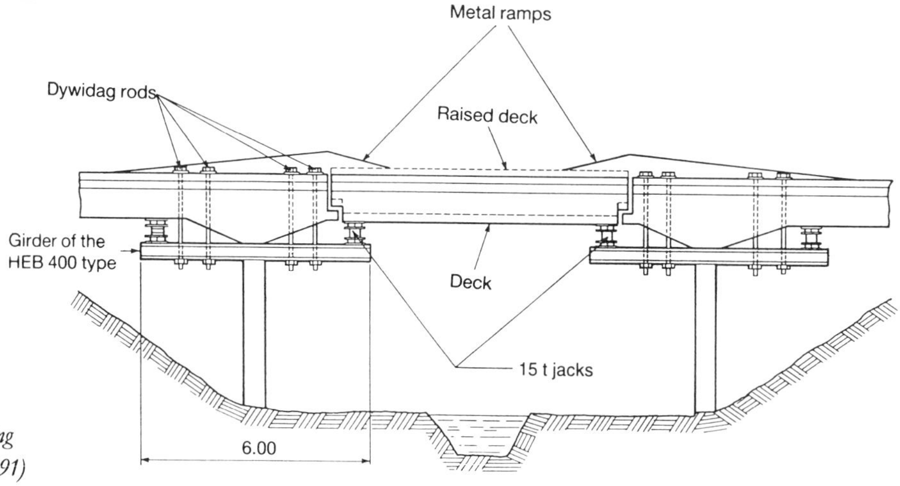

Fig. 2.10. Equipment for lifting suspended span (PIARC, 1991)

ules. With the maximum height 2.3 m and ramp gradients of 6%, it carried 70% of the passenger car capacity of the adjacent lane.

2.6.6 New Mikuni Great Bridge, Route 17, Japan—PIARC (1991) p 204, Nagata and Tatsushita

Built in 1963, the two lane 142 m span steel Langer truss has a 170 mm composite reinforced concrete deck slab. After 26 years the top 70 mm of the concrete had deteriorated badly, exposing the reinforcing steel. It was estimated that replacement of the deck by casting in place would require traffic control for 140 days. However, this was reduced to 70 days by the use of factory precast concrete deck units 2 m long by 3.8 m wide which were installed immediately after the demolition and preparation of the steel beams and the welding of shear studs. The work progressed at the rate of 6 m of one lane per day, the limits being set by saw-cuts 10 cm deep. Reinforcing steel across longitudinal joints was welded and lapped across the transverse joints. Distribution steel was fixed and the joints were completed with ultra-rapid hardening mortar which allowed opening to traffic after three to four hours curing and covering with temporary steel plates.

2.6.7 Deerbarn Bridge, A3 London–Portsmouth Trunk Road—PIARC (1991) p 255, Coxon

Built in 1934 to carry Guildford By-Pass over a railway, the three span reinforced concrete beam and slab deck is supported on reinforced concrete trestles and skeletal abutments. Prior to a widening scheme, a detailed investigation showed that cover to reinforcement varied from 0 to 75 mm and carbonation to depths of 20 mm with consequent excessive spalling. As the chloride content was insignificant and main reinforcement was not significantly reduced in section, it was decided to repair the underside using a polymer-modified cementitious mortar system. In the course of repairs, extensive corrosion of the beam stirrups was discovered and a new deck was constructed.

Access to the railway span was restricted to three consecutive weekends from 00.55 on Saturday to 05.25 on Monday with British Rail providing an alternative bus service. To reduce live loading and vibration, the road traffic was restricted to 40 mph during three sessions as follows.

- Contraflow northbound carriageway-repair east side.
- Contraflow southbound carriageway-repair west side.
- Offside lane closures-repair central section.

2.6.8 Crescent Bridge, A6 Trunk Road, Salford—PIARC (1991) p 293, Talbot, Griffiths and Lowe

Constructed in 1965, the 235 m long viaduct deck of precast pretensioned inverted T-beams with solid infill was supported on 21 sets of reinforced concrete columns and crossheads. Deterioration of the substructure was causing concern by 1980 and a detailed investigation concluded that the problem was due to calcium chloride having been used as an accelerator. Further monitoring in 1987 showed that cracking had increased by 40% since 1981 and some urgent repairs were necessary to maintain structural integrity. The trouble was attributed to the high porosity and low strength, 21 N/mm^2, of the concrete combined with general chloride concentrations of 2 to 5% by weight of cement. By contrast, the superstructure was in very good condition. As the columns and crossheads were beyond repair, it was decided to replace them completely, which involved temporary propping of the superstructure from the pile caps with military trestling. The crossheads were cut out in sections of about 7 t using a 1.5 mm water jet at 87.5 N/mm^2. The columns were quickly demolished with a hydraulic breaker attached to a mini-excavator. Starter bars were grouted into the pile caps and steelfixing, shuttering and pumping concrete to the columns caused no major problems. Plasticizer admixture was used for the crosshead. Traffic was unaffected throughout the whole of the works.

2.6.9 Seven motorway bridges in Cheshire—PIARC (1991) p 305, Washburn

The bridges are three span simply supported concrete decks on reinforced concrete trestles carrying the 33 m wide motorway M6 over roads, railways and canals. Deicing salt caused damage to sill beams and columns under leaking deck joints and the edges of in-situ concrete decks due to inadequate drainage of the central reserve. All defective and contaminated concrete was removed to a depth of 20 mm behind the reinforcement which was cleaned and augmented where required before making good with high strength concrete. Analysis showed that the relief of load required during the repairs could be achieved by prohibiting all traffic, including works traffic, from areas within one lane width of the repairs. The bridge repairs would have required two 13-week contracts but, by including this work with motorway repair and resurfacing work which would have needed a separate 20-week contract, the total period for three hybrid contracts was 33 weeks. This benefit was due to the dual use of costly contraflow systems needed for the particular combination of lengths of the motorway works and locations of the bridge works.

2.6.10 Alno Bridge, Sweden—PIARC (1991) p 333, Von Olnhausen

Built in 1964, the three high-level cantilevered prestressed concrete box-girder spans of 134 m provide 40 m clearance for navigation. Inspection in 1986 discovered cracking 45 m long and up to 1.8 mm wide separating the bottom slab from the southern web. This was deemed serious but not catastrophic. Traffic was restricted to the northern lane and lorries up to 25 t gross were permitted singly at night. Shear strengthening was provided by drilling 46 mm diameter holes and installing 32 mm diameter Dywidag prestressing bars throughout the webs at 900 mm centres and the bottom slab at 1800 mm centres. Restricted traffic was allowed to run during the six months duration of the work in the summer of 1987.

2.6.11 Djuro Bridge, Sweden—PIARC (1991) p 339, Von Olnhausen

Built in 1960–62, the 180 m span through-type arch has 20 m span continuous T-beam viaduct approaches giving an overall length of 350 m. In 1982 deicing salt damage to the deck slab was thoroughly investigated and remedial works were prescribed as follows:

- cover replacement to upper surface of bars
- concrete replacement around bars
- replacement of whole slab.

Rehabilitation required support from the ground. Water jetting was used to remove defective concrete. A study showed that vibration velocities up to 5 mm/s did not damage fresh concrete. Vehicles exceeding 4 t gross were prohibited during days of casting concrete.

2.6.12 High Bridge, Baden–Wettingen, Switzerland—PIARC (1991) p 351, Canton Aargau

The 240 m long reinforced concrete structure was built across the Limmat Valley in 1925–26, crossing the river with a 72 m span arch. After 50 years, considerable damage was detected to services under the footways and to the structural concrete. Rehabilitation included provision of a new deck slab, 30% wider than the original one. By phasing the work on a half-width at a time it was possible to maintain two lanes of traffic and a 2 m wide footway throughout. Comprehensive measurements showed that the vibration did not affect the setting of the concrete.

2.6.13 Structure X6, Egerkingen, Switzerland—PIARC (1991) p 385, Canton Solothurn

Built in 1966, the two 50° skew slab decks carry four lanes southbound and five lanes northbound of Highway N2 over a Cantonal road. The slabs suffered damage due to ingress of salt water to their adjoining edges and at their supports due to leaking expansion joints. The bearings and abutment sill beams were also badly corroded. It was decided to make a clear opening longitudinally between the two slabs, to remove defective concrete from the support regions and make the deck slabs monolithic with the abutment walls. Rehabilitation was carried out in 1989–90, keeping two lanes of the N2 open in each direction throughout, except for two closures of one night each when traffic was diverted to the opposite carriageway. Traffic on the Cantonal road was controlled by traffic lights and a height limit of 4.3 m was applied.

2.7 Alkali-silica reaction

2.7.1 Marsh Mills Viaduct, Plymouth—*New Civil Engineer* (1985 and 1986) and Department of Transport (1993)

The viaduct was constructed in 1969–70 and, in the late 1970s, cracking developed. This was thoroughly investigated and in 1980 its cause was diagnosed as expansion due to alkali-silica reaction. It severely affected piles and pile caps with cracks up to 20 mm wide and heavy spalling. It was also identified in piers, crossheads, deck slabs and abutments. Following careful analysis of the structure, the carriageway was reduced from three lanes to two and load restrictions were applied. Steel props were installed to transfer load direct to the piles where pile caps were reactive. Also various measures were taken to reduce the incursion of water into the structure to reduce the development rate of the reaction. The measures were waterproofing, improved deck drainage and shielding of exposed concrete copings by the use of preformed panels. The condition of the structure is being monitored until its replacement in 1994–95.

2.7.2 Railway bridge piers, Canada—Houde, Lacroix and Morneau (1987)

The 44 simply supported plate girder spans of 22 m crossing water were supported on 1.8 × 3.6 m concrete piers. Tests showed that all of the concrete above the water-line had to be replaced. While the damaged concrete was removed and replaced, the superstructure was carried by steel columns embedded 5 to 10 m deep in coring holes in the piers below water-level.

2.7.3 Hanshin Expressway, Japan—Imai, Yamasaki, Maehara and Miyagawa (1987)

The structural behaviour of deteriorated concrete is described with the preferred repair method using synthetic resin. Natural and accelerated exposure trials were conducted on concrete specimens treated with the following materials:

- ☐ epoxy resin coating
- ☐ polymer cement paste lining
- ☐ silane monomer impregnation
- ☐ silane oligomer impregnation.

Field repairs to affected piers showed that both corrosion of reinforcement and expansion of concrete due to alkali-silica reaction were effectively controlled by filling cracks by pressure injection, followed by either an epoxy coating or silane impregnation.

2.7.4 Pell Street I/C, Port Elizabeth, South Africa—Vanderstraeten (1987)

The investigations to determine the presence and effects of alkali-silica reaction are outlined. The findings and the adopted remedial measures are described, in particular the method of encapsulating the affected pile caps with reinforced and prestressed concrete.

2.7.5 Bridges in Northern France—Godart, Fasseau and Michel (1989)

The information in Table 2.1 was presented.

Table 2.1. Inventory of bridges with ASR

Department	Number of bridges built since 1970	Number of bridges affected by ASR	%
Pas de Calais	290	20	7
Nord	400	100	25
Somme	80	0	0
Aisne	90	3	3

Monitoring was in three modes:

- ☐ local and global cracking using a crack meter with the results given in Table 2.2;
- ☐ global deformation;
- ☐ residual expansion of core.

Table 2.2. Classification of cracking.

Cracking indices (mean total opening in mm/m)	Extent of damage
0 – 0.5	Negligible
0.5 – 1	Low
1 – 2	Moderate
2 – 5	High
5 – 10	Very high
> 10	Considerable

2.8 Fire damage

2.8.1 Bristol Parkway Viaduct—Wolley (1972)

Two hundred EPS void formers, 900 mm in diameter and 2.4 m long, were stored under a 32 m span of a slip road viaduct. The span was a prestressed slab 1220 mm deep with 915 mm diameter voids not formed with EPS. The material was set alight, possibly by some children, and burned for about 15 to 20 minutes releasing an estimated 100 000 MJ (enough to run a domestic central heating system for a year). The temperature of the smoke and fumes was probably between 800 and 1200°C. Although the fire had caused spalling to a depth of 40 mm over much of the soffit, exposing the mild steel transverse reinforcement, load testing showed that the span was still structurally sound and the damage was repaired by guniting. Site precautions were recommended.

2.8.2 Motorway overbridge—*New Civil Engineer* (1982)

A lorry parked beneath a bridge over the M6 motorway caught fire and caused the steel deck beams to sag at least 150 mm. Firemen fighting the blaze had to contend with falling chunks of spalling deck concrete. The bridge bearings were damaged and reinforcement exposed on the deck soffit and a pier face.

2.8.3 Keele Service Station Bridge—*Construction News* (1984)

A fire destroyed the section of the service station building on the bridge over the M6 motorway at Keele, Staffordshire. There was no significant structural damage to the bridge.

2.8.4 Littlebrook Bridge, Dartford—*New Civil Engineer* (1984)

EPS soffit formwork caught fire on the newly built Littlebrook Bridge over the southern approach to the Dartford Tunnel. It was thought to have flashed across the 30 m length of the bridge in less than 15 seconds and to have lasted some 15 minutes. The fire occurred at 22.15 while the night shift was working. Vandalism was ruled out by the police. Oxyacetylene cutting had been in progress that evening but not at the time of the fire. The contractor investigated possible causes associated with hot wire cutting equipment used to remove the EPS soffit forms although the supplier ruled out any malfunction in the system used every day. Immediately after the fire, temporary support was installed to safeguard the structure. Although there was some spalling on the eight prestressed concrete bridge beams they were believed to be structurally sound.

2.8.5 Gravelly Hill Interchange—*New Civil Engineer* (1990 and 1991)

The driver died when a tanker travelling west along the A38 with 33 000 litres of petrol crashed into the substructure of a slip road viaduct at 07.25 on Sunday

26 August 1990. The ensuing fire caused extensive damage to a main 46 m span and adjacent spans at two levels and two columns. About a quarter of the main span and spherical pot bearings were damaged. On investigation it was reported that the temperature had reached 600°C and 75% of the affected areas would have to be replaced. To keep the traffic flowing on the A38 and reopen the motorway interchange roads as soon as possible, three options were considered:

- repair the deck under dead load,
- repair the deck under live load,
- complete replacement.

After a sensitivity analysis of some of the unknown factors, the Department of Transport decided upon total replacement of the damaged structure and the work was completed exactly one year after the accident.

2.9 Impact damage

2.9.1 Kempley Road Bridge—*New Civil Engineer* (1978)

The continuous six span bridge was built in 1960, before the introduction of an impact load design standard, to carry a minor road over the M50 motorway. An articulated lorry left the motorway and swept away a slender reinforced concrete portal framed support. Two spans of the bridge collapsed, killing the driver. Other spans and supports were displaced and the structure was beyond repair.

2.9.2 Common Lane Bridge—Gledhill and Jones (1981)

The bridge deck, built in the early 1960s, comprised precast pretensioned concrete inverted T-beams with post-tensioned diaphragms and a reinforced concrete deck slab. It was struck by a heavy item of plant carried on a low-loader, resulting in severe local damage to the bottom flange of one edge beam. The repair consisted of additional prestressing wires immediately above the damaged flange with new anchorage blocks behind the ends of the original transverse diaphragms reinforced with stirrups housed in slots cut carefully through the deck slab. The supplementary prestressing was then hidden from view by reinstated fascia panels, as originally designed to mask the end blocks of the transverse diaphragms.

2.9.3 Cattlelaithe Bridge—Gledhill and Jones (1981)

A similar impact on this accommodation bridge over the M62 motorway resulted in the loss of some strands and severe web cracking. This could not be repaired by the introduction of limited supplementary prestress as there were no diaphragms in the span with which to locate anchorages. A replacement edge beam was post-tensioned and integrated into the existing deck at the end diaphragms.

2.9.4 Dolphin Footbridge—Gledhill and Jones (1981)

This four span prestressed concrete U-section footbridge over the M62 motorway was also designed to carry some public utility services and, when it was hit by an item of plant on a low-loader, two 150 mm diameter steel sleeves provided the major restraint preventing its collapse onto the carriageway. The spans were supported alternately by dowelled and rubber bearings. One end of the impacted span was moved across the pier which had tilted and failed at its root while the main span beam soffit and the web of a side span beam were cracked. The beams were repaired by epoxy grouting the cracks and prestressing the soffit and side walls with Macalloy bars. The pier was relieved of load, plumbed and its root secured by a prestressed and reinforced concrete collar.

2.9.5 Prestressed suspension bridge—Kreps-Heyndrikx and Callens (1981)

Built in 1957 to a unique design by Professor D. Vandepitte of the University of Gent, the concrete deck slab of this bridge was prestressed by the tension in the suspension cables. Each of the two cables passed continuously throughout the deck and over the tops of two reinforced concrete portals. The initial prestress was 7000 kN and the vertical reaction in the portal column was 4600 kN. In 1976 a lorry with a high load struck the horizontal member of one portal, causing local damage and inward displacement of the columns which developed plastic hinges. Fortunately the lorry stopped before reaching the second portal which would have been catastrophic.

The repair consisted of restoring the columns to their initial positions, replacing the damaged concrete while maintaining the reaction in order to support the suspension cables. This was achieved by the use of temporary steel trestles and struts to transfer load from the upper sections of the columns to the piers. This relieved the plastic hinge zones and the crossbeam of load while they were then reconcreted and gained the required strength.

2.10 References

AL-BABTAIN I. and ABBAS A. M. (1991). Maintenance and repair of highway concrete bridges: a case study. *Strategic Highway Research Program and Traffic Safety on two Continents*. Gothenburg. 18–20 Sept.

AL-MANDIL M. Y., AZAD A. K., BALUCH M. H., PEARSON-KIRK D. and SHARIF A. M. (1989). Punching shear failure of concrete girder-slab type bridge decks in Saudi Arabia. *Proc. Int. Conf. on Structural Faults and Repair*. Kensington, London, 27–29 June, **1**.

AMBROSE K. (1985). Chloride contamination of Camsley Lane Viaduct. *Construction Repairs and Maintenance*, Sept.

ANDERSON G. (1990). Chloride extraction and realkalisation of concrete. *Hong Kong Contractor*. July-Aug.

ANGERER R. A. (1993). Zuoz local authority. Private communication.

ARCHIER J. (1989). Repair of the Veauche Bridge over the River Loire (in French). *Symp. on the Durability of Structures*. Lisbon. International Association for Bridge and Structural Engineering, 937–941.

BARFOOT J. (1985). Repairing viaduct crossheads at Coventry. *Construction Repairs and Maintenance*. July, 15–17.

BILLINGTON D. P. (1979). *Robert Maillart's bridges—the art of engineering*. Princeton University Press, New Jersey, USA, and Guildford, UK.

BOAM K. J. and UNWIN J. (1990). The Midland Links—a maintenance strategy. *Institution of Highways and Transportation National Workshop: New Concepts for the Management of Highway Structures*. Leamington Spa, 20 Mar.

BORDERIE P. (1981). The repair of the Fives prestressed road bridge. *Proc. Int. Conf. on Inspection, Maintenance and Repair of Road and Railway Bridges*. 13–17 Apr., Bruxelles–Paris, 273–278.

BRIEN A. T. (1990). Concrete repairs in the north of Scotland. *British-Swedish Seminar on Concrete Repairs*. Pubn P-90:11. Chalmers University of Technology, Goteborg.

CADY P. D. and WEYERS R. E. (1987). Deterioration of concrete bridge decks from corrosion of reinforcing steel. *Concrete International: Design and Construction*. Jan., 15–20.

CAVALIER P. G. and VASSIE P. R. (1981). Investigation and repair of reinforcement corrosion in a bridge deck. Paper 8449. *Proc. Instn Civ. Engrs*, Part 1, **70**, Aug., 461–480.

CAVALIER P. G. and VASSIE P. R. (1982). Investigation and repair of reinforcement corrosion in a bridge deck. Discussion 8449. *Proc. Instn Civ. Engrs*, Part 1, **72**, Aug., 401–419.

CHALKLEY B. (1989). Corrosion on the Japanese Express. *Construction and Repair International*, June, 32–33.

COGSWELL G. and HERBERT A. P. (1991). A25 Woodbridge (Old) Guildford Bridge. *J. Instn Highw. and Transpn*, Sept., 7 *et seq.*

CONSTRUCTION NEWS (1991). Re-forging old links. (Midland Links). Crossbeam repairs, cathodic protection and the Tyburn Road fire. *Construction News*, 17 Jan., 12.

CONSTRUCTION REPAIR INTERNATIONAL (1989). Concrete crossheads replaced on M56 Weaver Viaduct. *Construction Repair International*, Jan./Feb., 13.

CRERER J. (1993). Private communication for the Engineer to Tay Road Bridge Joint Board.

DAHINTER K. (1990). Inspection and strength evaluation of concrete highway bridges in Czechoslovakia. Illustrative examples. *Proc. 1st Int. Conf. on Bridge Management*. University of Surrey, Guildford. 28–30 Mar., 386–393.

DARBY J. J. (1992). Maintenance—a County viewpoint. *Durability of Concrete Structures Symposium*. Institution of Highways and Transportation, South-Western Branch, Taunton, 12 Nov. Unpublished.

DARBY J. J. (1993). Private communication.

DEPARTMENT of TRANSPORT (1985). *Midland Links Viaducts*. DoT Interim Report, July.

DEPARTMENT of TRANSPORT (1988). *Repair and maintenance of the Midland Links Viaducts*. Working Party Report.

DEPARTMENT of TRANSPORT (1993). Private communications.

DIXON J. (1988). Construction repair—lifeline or liability? *Concrete Repair Conf.*, London Tara Hotel, 16 Mar., Palladian Publications Ltd.

DJURDJEVIC M. and PAKVOR A. (1985). Design for the repair of Pancevo Bridge and IJKL expansion joint. *Zbornik Simpozijum*, nm, 45–50 (in Serbian).

DJURDJEVIC M. and PAKVOR A. (1989). Causes of damage and rehabilitation of the Pancevo Bridge, Belgrade. *Symp. on the Durability of Structures*. Lisbon. International Association for Bridge and Structural Engineering, 671–676.

ESTIE J. B. (1993). Technical Services Consultancy. Northumberland County Council. Private communication.

EVANS R. A. and LEE D. M. B. (1981). Corrosion fight on Pelham Bridge, Lincoln. *The Highway Engineer*, **28**, No. 10, Oct., 16–21.

FAVRE R. (1969). Die Erneuerung von zwei Maillart-Brücken. *Schweizerische Bauzeitung*, No. 17, 313–319.

FREARSON J. P. H., MARSHALL G. R. D. and HAMILTON P. A. (1987). The Manama–Sitra Causeway bridges—the investigation and assessment of concrete condition. *Proc. 2nd Int. Conf. on Deterioration and Repair of Reinforced Concrete in the Arabian Gulf*. Bahrain Society of Engineers.

FREEH W. S. (1984). Ohio Turnpike, Cuyahoga River Bridge Rehabilitation. *Transportation Research Record*, **1**, No. 950, 173–176.

FUZIER J. P. (1990). A. S. Nowak (ed.). Reinforcement of N9 motorway bridges over the Lutrive Valley near Lausanne—Switzerland. *Bridge evaluation, repair and rehabilitation*. Nato ASI Series E: Applied Sciences, Vol. 187, ISBN 0 7923 0999 5, 567–577.

GLEDHILL C. and JONES C. J. F. P. (1981). The repair of bridges damaged by vehicular impact. *Proc. Int. Conf. on Inspection, Maintenance and Repair of Road and Railway Bridges*. 13–17 Apr., Bruxelles-Paris, 339–344.

GODART B., FASSEAEU P. and MICHEL M. (1989). Diagnosis and monitoring of concrete bridges damaged by AAR in Northern France. *Proc. 9th Int. Conf. on AAR in Concrete*. Queen Elizabeth II International Conference Centre, London, 27–31 July, 368 *et seq*.

GOGIC D. and MOZINA B. (1985). Repair of the road bridge across the D-T-D Canal on the Novi Sad–Temerin Road (in Serbian). *Zbornik Simpozijum*, nm, 51–56.

GOLDSBY I. F. (1990). The strengthening and repair of Rakewood Bridge. *J. Instn Highw. and Transpn*, Apr., 39–43.

GROZIER A. C. (1974). Strengthening a 50-year old viaduct. *Concrete*, July, 22–25.

GUARDIAN (1991). Span of history. *The Guardian*, 14 Jan.

HOUDE J., LACROIX P. and MORNEAU M. (1987). Rehabilitation of railway bridge piers heavily damaged by alkali-aggregate reaction. *Proc. 7th Int. Conf. on Concrete Alkali-aggregate Reactions*. Ottawa, 1986, 163–167.

IMAI H., YAMASAKI T., MAEHARA H. and MIYAGAWA T. (1987). The deterioration by alkali-silica reaction of Hanshin Expressway concrete structures. *Proc. 7th Int. Conf. on Concrete Alkali-aggregate Reactions*. Ottawa, 1986, 131–135.

ISECKE B. (1990). Long term behaviour of materials in a prestressed concrete bridge. *Proc. 3rd Int. Conf. on Corrosion of Reinforcement in Concrete Construction*. Wishaw, Warwickshire, 21–24 May. Published for the Society of Chemical Industry by Elsevier Applied Science, 142.

JACKSON C. (1989). Motorway upgrading project completed ahead of schedule. (Barton High Level Bridge). *J. Instn Highw. and Transpn*, Feb., 5–6.

JOHNSON P. (1993). G. Maunsell and Partners, Consulting Engineers to the Department of Transport. Private communication.

JOVOVIC O. (1989). Repair of a pedestrian bridge of prestressed concrete. *Proc. Int. Conf.* on Structural Faults and Repair, Kensington, London. 27–29 June, **1**, 335–338.

KREPS-HEYNDRIKX Y. and CALLENS M. (1981). Travaux de réparation d'un pont-route en béton précontraint. *Proc. Int. Conf. on Inspection, Maintenance and Repair of Road and Railway Bridges*. 13–17 Apr., Bruxelles-Paris, 363–365.

KREPS-HEYNDRIKX Y. and CALLENS M. (1981). Travaux de réparation d'un pont suspendu précontraint, heurté par un véhicule. *Proc. Int. Conf. on Inspection, Maintenance and Repair of Road and Railway Bridges*. 13–17 Apr., Bruxelles-Paris, 366–370

LEON R. T., FRENCH C. W. and OLSON S. A. (1989). Prestressed concrete girders after 20 years service. *Symp. on the Durability of Structures*. Lisbon. International Association for Bridge and Structural Engineering.

LUTZ J. G. (1990). A. S. Nowak (ed.). Rehabilitation of the Harry W. Nice Memorial Bridge. *Bridge evaluation, repair and rehabilitation*. Nato ASI Series E: Applied Sciences, Vol. 187, ISBN 0 7923 0999 5, 589–601.

MCCARTER D. B. (1993). Highways Department. Norfolk County Council. Private communication.

MACCRAITH S. (1990). Performance of an 80 year old reinforced concrete bridge in an extreme environment. *Proc. 3rd Int. Conf. on Corrosion of Reinforcement in Concrete Construction*. Wishaw, Warwickshire, 21–24 May. Published for the Society of Chemical Industry by Elsevier Applied Science, 188.

MACMILLAN G. L. and TREADAWAY K. W. J. (1990). An exposure trial of concrete durability in Arabian Gulf conditions. *Proc. 3rd Int. Conf. on Corrosion of Reinforcement in Concrete Construction*. Wishaw, Warwickshire, 21–24 May. Published for the Society of Chemical Industry by Elsevier Applied Science, 115.

MARSHALL G. R. D. and CHANNA R. (1987). The Manama–Sitra Causeway bridges—replace or repair; the engineering and economic assessments. *Proc. 2nd Int. Conf. on Deterioration and Repair of Reinforced Concrete in the Arabian Gulf*. Bahrain Society of Engineers.

MAXWELL J. W. S. (1985). Problem identification and remedial measures on the Tay Road Bridge. *Proc. 2nd Int. Conf. on Structural Faults and Repair*, Institution of Civil Engineers, London, 30 Apr.–2 May.

MILANOV S. (1988). Repair of the bridge across the D-T-D Canal on the M24 main road near Melenac (in Serbian). *Put i Saobracaj*, **34**, No. 5/6, 28–32.

MILLER D. O. and BECKMAN R. D. (1984). Renovation of the Third Avenue Bridge in Minneapolis. *Transportation Research Record*, **1**, No. 950, 150–156.

NEW CIVIL ENGINEER (1978). Easing the tension over Braidley Road Bridge. *New Civil Engineer*, 31 Aug. 1978, 12–13.

NEW CIVIL ENGINEER (1978). Driver killed as lorry brings down bridge. *New Civil Engineer*, 14 Dec. 1978, 6–7.

NEW CIVIL ENGINEER (1980). Bridges. Unbonded tendons: why the fall from favour? *New Civil Engineer*, 24 Jan. 1980, 20–24.

NEW CIVIL ENGINEER (1980). Cracking clampdown gives Guildford respite. *New Civil Engineer*, 7 Feb. 1980, 24–25.

NEW CIVIL ENGINEER (1982). News. Buckled beams. *New Civil Engineer*, 24 June 1982, 5.

NEW CIVIL ENGINEER (1984). Column top recast on Midland Links. *New Civil Engineer*, 1 Mar. 1984, 7.

NEW CIVIL ENGINEER (1984). Corrosion threatens modern bridges. (Taf Fawr Bridge, Thelwall Viaduct and others). *New Civil Engineer*, 23/30 Aug. 1984, 4.

NEW CIVIL ENGINEER (1984). Cover story. Corrosion takes its toll on Tay. *New Civil Engineer*, 20 Sept. 1984, 33–35.

NEW CIVIL ENGINEER (1984). Bridge fire probe starts. *New Civil Engineer*, 15 Nov. 1984, 5.

NEW CIVIL ENGINEER (1985). Ring road trials may herald £4m repairs. (Midland Links). *New Civil Engineer*, 21 Feb. 1985, 4 *et seq*.

NEW CIVIL ENGINEER (1985). Cover story. Marsh Mills Viaduct. *New Civil Engineer*, 24 Apr. 1985.

NEW CIVIL ENGINEER (1985). Repairs vital on AAR-hit Midland Links. *New Civil Engineer*, 1 Aug. 1985, 4 *et seq*.

New Civil Engineer (1985). Cover story. Taff temporary takes up traffic. *New Civil Engineer,* 3 Oct. 1985, 34–35.

New Civil Engineer (1985). Welsh bridge failure starts tendon scare. *New Civil Engineer,* 12 Dec. 1985, 4–5.

New Civil Engineer (1986). Report pinpoints crack attack. *New Civil Engineer,* 12 June 1986, 24–25.

New Civil Engineer (1986). Manholes abetted A1 viaduct corrosion. News. *New Civil Engineer*, 17 July 1986, 6.

New Civil Engineer (1986). Bridges. Traffic makes way for Barton widening. *New Civil Engineer*, 4 Sept. 1986, 32–36.

New Civil Engineer (1987). Midland Links analysis offers full repair range. *New Civil Engineer*, 9 Apr. 1987, 5 *et seq.*

New Civil Engineer (1987). M1 bridge bars checked for stress. *New Civil Engineer,* 16 Apr. 1987, 9.

New Civil Engineer (1988). Thelwall duplication key to M6 widening. *New Civil Engineer,* 14 Jan. 1988, 8.

New Civil Engineer (1988). Viaduct revamp. *New Civil Engineer,* 18 Aug. 1988, 14–17.

New Civil Engineer (1988). Bridges. Gantry ready to repair. (Tay Road Bridge). *New Civil Engineer,* 17 Nov. 1988, 20–22.

New Civil Engineer (1989). Record repair planned for Midland Links. *New Civil Engineer,* 23 Feb. 1989, 5.

New Civil Engineer (1989). Tay Bridge to be cathodically protected. *New Civil Engineer,* 6 July 1989, 11.

New Civil Engineer (1990). Tendon corrosion closes M1 viaduct. *New Civil Engineer*, 3 May 1990, 10.

New Civil Engineer (1990). Spaghetti junction lifts off £400m repairs, DoT must steer clear of false economy. *New Civil Engineer,* 28 June 1990, 19.

New Civil Engineer (1990). Gravelly Hill fire. *New Civil Engineer,* 6 Sept. 1990.

New Civil Engineer (1990). Gravelly Hill rebuild. *New Civil Engineer,* 13 Dec. 1990.

New Civil Engineer (1991). Crosshead repairs promise M4 chaos. *New Civil Engineer,* 18 July 1991, 8.

New Civil Engineer (1991). Wardley Hall Viaduct piers replaced. *New Civil Engineer,* 1 Aug. 1991.

New Civil Engineer (1991). Running repair. (Midland Links—Gravelly Hill fire damage). *New Civil Engineer,* 15 Aug. 1991, 17–19.

New Civil Engineer (1991). Crescent forerunner. *New Civil Engineer,* 19 Sept. 1991, 19.

New Civil Engineer (1991). Desalination set to cut repair bills. *New Civil Engineer,* 14 Nov. 1991, 6.

New Civil Engineer (1992). Bridge flex worries delay M5 beam break out. *New Civil Engineer,* 26 Mar. 1992, 7.

New Civil Engineer (1992). Viaduct span beyond repair. *New Civil Engineer*, 30 Apr. 1992, 8.

New Civil Engineer (1994). Tendon corrosion closes 'cursed' A3 bridge. *New Civil Engineer*, 30 June 1994, 4–5.

Olson S. A., French C.W. and Leon R. T. (1990). A. S. Nowak (ed.)., Prestressed concrete girders after 20 years service. *Bridge evaluation, repair and rehabilitation.* Nato ASI Series E: Applied Sciences, Vol. 187, ISBN 0 7923 0999 5, 391–403.

Pakvor A., Tarana J. and Darijevic Z. (1983). State-of-the-art and design of road bridge repair across the Danube–Tisza–Danube Canal near Srpski Miletec (in Serbian). *Zbornic, vii Kongres Saveza Drustava Gradjevinskih Konstrukteru*, nm, 257 263.

PAKVOR A. and DJURDJEVIC M.(1989). Causes of damage and rehabilitation of the Pancevo Bridge, Belgrade. *Symp. on the Durability of Structures.* Lisbon. International Association for Bridge and Structural Engineering, 671–676.

PIARC TECHNICAL COMMITTEE ON BRIDGES (1991). *Repair of bridges under traffic.* Permanent International Association of Road Congresses. 27, rue Guenegaud—75006 Paris.

ROBINS P. J. (1990). Fibre shotcrete repair applications in the UK. *British-Swedish Seminar on Concrete Repairs.* Pubn P-90:11. Chalmers University of Technology, Goteborg.

RYELL J., MANNING D. G. and OWENS M. (1984). *Durable highway structures—recent developments in Ontario.* Ministry of Transportation and Communications, Ontario.

SCHNEPPENDAHL G. (ed.) (1984). *Concrete repair and protection.* Sika AG, Zürich.

SHIELDS J. (1994) Private communication for the Engineer to Tay Road Bridge Joint Board.

SMITH B. J. and STRATTON F.W. (1988). *Bridge deck hollow plane repair using injected epoxy.* Final Report. Kansas Department of Transportation, USA.

SRAM S. (Professor) (1985). Experience in the maintenance of the Bridge to the Island of Pag (in Croatian). *Zbornik Simpozijum*, nm, 195–202.

SRAM S. (Professor) (1987). The Pag Bridge maintenance and repair problems (in Croatian). *Cavtat, viii Kongres Saveza Drustava Gradjevinskih Konstruktera*, 7–10 Apr., Proceedings, Vol. m, 229–234.

STANLEY C. C. (1979). *Highlights in the history of concrete.* Cement and Concrete Association (British Cement Association).

STRATFULL R. F. and SEIM C. (1979). *Corrosion of steel piles at the Richmond–San Rafael Bridge.* PB298823. FHA Sacramento, CA.

TASSIOS T. and ALIGIZAKI K. (1990). A. S. Nowak (ed.). Evaluation of an old prestressed concrete bridge. *Bridge evaluation, repair and rehabilitation.* Nato ASI Series E: Applied Sciences, Vol. 187, ISBN 0 7923 0999 5, 435–443.

THODAY N. J., JONES D. and SIMPSON A. (1993). Investigation, repair and strengthening of Wardley Hall Bridge. *Proc. 2nd Int. Conf. on Bridge Management.* University of Surrey, Guildford, 18–21 Apr.

THOMPSON D. M. (1989). Effectiveness of repair of reinforcement corrosion in a motorway viaduct. *Proc. 3rd Int. Conf. on Deterioration and Repair of Reinforced Concrete in the Arabian Gulf.* Bahrain Society of Engineers.

TORK A. (1985). Investigation of prestressed concrete box beams of an elevated expressway. *Canadian J. Civ. Engng*, Mar., **12**, No. 1, 1–11.

VANDERSTRAETEN A. T. (1987). Repair of alkali-aggregate reaction damage to the motorway structures of the Pell Street Interchange, Port Elizabeth, Republic of South Africa. *Proc. 7th Int. Conf. on Concrete Alkali-aggregate Reactions.* Ottawa, 1986, 194–198.

VAYSBURD A. M. (1990). Rehabilitation of an elevated roadway bridge. *Concrete International: Design and Construction*, **12**, No. 9, 45–50.

WATTERS A. (1991). Cathodic protection of Tay Road Bridge substructure. *Construction Maintenance and Repair*, Jan./Feb., 38–45.

WEYERS R. E. and CADY P. D. (1987). Deterioration of concrete bridge decks from corrosion of reinforcing steel. *Concrete International: Design and Construction*, **9**, No. 1, 15–20.

WOLLEY C.W. (1972). The motorway that caught fire. *Concrete*, Sept., **6**, No. 9, 32–35.

WOODWARD R. J. and LOE J. A. (1981). The inspection and maintenance of post-tensioned concrete bridges. *Proc. Int. Conf. on Inspection, Maintenance and Repair of Road and Railway Bridges*, 13–17 Apr., Bruxelles-Paris, 265–269.

WOODWARD R. J. (1981). *Conditions within ducts in post-tensioned prestressed concrete bridges.* TRRL Laboratory Report 980.

WOODWARD R. J. and WILLIAMS F. W. (1989). Collapse of Ynys-y-Gwas Bridge, West Glamorgan. Paper 9357. *Proc. Instn Civ. Engrs*, Part 1, Aug., 635–670.
ZICHNER T. (1989). Zweckässigkeit und Umfang von Instandsetzung. Suitableness and extent of rehabilitation (in German). *Symp. on the Durability of Structures*. Lisbon. International Association for Bridge and Structural Engineering, 665–670.
ZICHNER T. (1993). Private communication.

3

Detection and monitoring

3.1 Routine inspections

3.1.1 Bridge management and inspections

The Standing Committee on Structural Safety (1977) reviewed the carefully organized inspection and maintenance of railway structures, planned from Board level down, using standardized procedures. SCOSS (1985) noted that, despite budgetary cuts in maintenance and renewals, the condition of railway structures gave no cause for immediate concern. They understood that the maintenance budget had been kept at about the same relatively low level, approximately 0.17% of the replacement value of the structures, for nine years. OECD (1981) collated highway bridge maintenance costs, ranging from 0.3 to 2.5%, excluding renewal costs. The large range was due to differences in estimating replacement costs and a number of other factors but the data were interpreted to infer that 0.5% was a reasonable average required to prevent progressive deterioration. (See also 1.1.2.)

The Department of Transport (1977) prescribed Superficial, General, Principal and Special Inspections for its highway structures. The Bridge Inspection Panel (1984), concerned mainly with General and Principal Inspections, outlined the essential actions required for effective inspection, assessing defects and reporting their severity and extent.

The Department of Transport (1986) proposed that Special Inspections should be considered to investigate cracks and other defects, to undertake appropriate testing and monitoring and to propose repairs. Pullar-Strecker (1987) gave useful guidance on the assessment and repair of corrosion-damaged concrete. The Department's later requirements (1988) included forms for the transmission of inspection information to the database. Thus the first signs of deterioration are normally picked up in General or Principal Inspections and the condition and repair history of each structure is monitored. The Author of this Review (1986) outlined the needs, aims and various aspects of bridge maintenance and Boyes (1990) presented an updated report of the bridge management procedures adopted by the DoT.

Several conferences have been devoted to the management of bridges, covering inspection, monitoring and data handling systems.

Proceedings of the First International Conference on Bridge Management have been edited by Harding and Parke (1990) and comprise 67 papers from 25 countries. The Second Conference was held on 18–21 April 1993.

Proceedings of the NATO Advanced Workshop: Bridge Evaluation, Repair and Rehabilitation edited by Nowac (1990) start with management, diagnostics, monitoring, evaluation and testing.

The latest OECD (1992) Road Research Report, Bridge Management, reviews the development of Bridge Management Systems, BMS, in OECD member countries.

3.1.2 Visual inspection

A General Inspection requires the visual examination of representative parts of the structure. Visible defects in reinforced concrete are likely to be cracking, scaling, spalling or staining.

A Principal Inspection requires close examination of all inspectable parts of the structure. This may need access and other equipment. For instance, during the first Principal Inspection of the Orwell Bridge, Suffolk, central faces of the twin boxes were inspected by Closed Circuit Television because they were inaccessible by the underbridge access gantry used for the other external surfaces at that time.

Barry (1989) described the application of abseiling for which the Health and Safety Executive had drafted guidance notes. An abseiler, having both hands free, can inspect, sound with a hammer, photograph, measure, drill, test or repair.

The cause of cracking must be diagnosed. A great many cracks have been caused in the early days after construction by the tensile stresses that result from the various restraints to shrinkage. The whole gamut of non-structural cracking has been described in Concrete Society Technical Report No. 22 (1992).

Controlled structural cracking is, of course, inherent in all reinforced concrete. The significance of cracks in relation to durability has been the subject of debate, such as that between Darwin, Manning, Hognestad, Beeby, Rice and Ghowrwal (1985). On the one hand, Beeby (1978) maintained that, in properly designed reinforced concrete, cracks are closely spaced and their width decreases from the nominal design width at the surface to zero before reaching the depth of cover to the reinforcement. Therefore such cracks do not provide a path for water and aggressive agents to produce corrosion of the steel; shrinkage cracks do not taper down to zero in that way and may be much the same width at the steel as at the surface. Neville (1981) considered that cracks in concrete may be capable of autogenous healing, thus preventing corrosion. These premises imply that corrosive agents, notably chloride ions, enter by diffusion or by non-structural cracks. Others have argued that corrosion is related to structural cracking as well. Borgard *et al.* (1989), Transportation Research Record No. 1211, pp 1–11, reported evidence, obtained by means of a scanning electron microscope, of chloride-induced corrosion in the vicinity of a crack but a complete absence of chloride ions only a few millimetres away. They claimed that this shows that chloride ions entered by the crack and not by diffusion.

However, the important distinction affecting repair is whether a crack is dormant or active. A tell-tale may be fixed across a crack; if it remains intact it confirms that the crack is dormant but if it breaks it does not necessarily prove that the crack is structural. A demec gauge or a microscope width gauge will provide more positive information. If monitoring shows no growth in the cracking, the remedy is simply to seal it against the ingress of water and any aggressive agents.

If, however, cracks are either propagating or widening, or repaired cracks are reopening, a geophysical or structural cause should be found and remedied before making the permanent repair. Map-cracking suggests alkali-silica reaction which can be diagnosed with the guidance of the Cement and Concrete Association (now BCA) Working Party Report (1988).

Scaling is usually due to frost action. Spalling or staining are usually signs that the reinforcement is corroding although corrosion can be active with no evidence on the surface of the concrete.

Defects are commonly due to more than one cause.

Colour photographs of defects are useful for comparison at future inspections. Visible signs may lead to a diagnosis or point to the need for further investigation.

Delamination near the surface can be detected by tapping with a light hammer and listening for a dull response.

3.2 Concrete testing

3.2.1 Cover depth

Depth of concrete cover is measured from the distortion by a reinforcing bar of a magnetic field generated at the surface by the instrument. The operation is completely non-destructive but usually requires knowledge of the bar diameter. There is a wide range of covermeters available. Bungey (1992a) suggested that their accuracy is unlikely to be better than ± 5 mm. While bars can be detected up to 200 mm deep, reliable cover estimates may only be possible up to 100 mm for single bars and the calibration should always be checked by drilling.

Alldred (1993) described the use of a covermeter for measuring bars of unknown diameter by a method based on ratios of signal strengths rather than differences in indicated covers.

3.2.2 Rebound hammer

This equipment, developed by Ernst Schmidt in Switzerland in the 1940s, was fully described by Kay (1992). It measures the surface hardness and may be useful in making a comparative study of cover concrete or locating weak areas but it relates only to a very limited depth below the surface and is affected by the surface texture and carbonation.

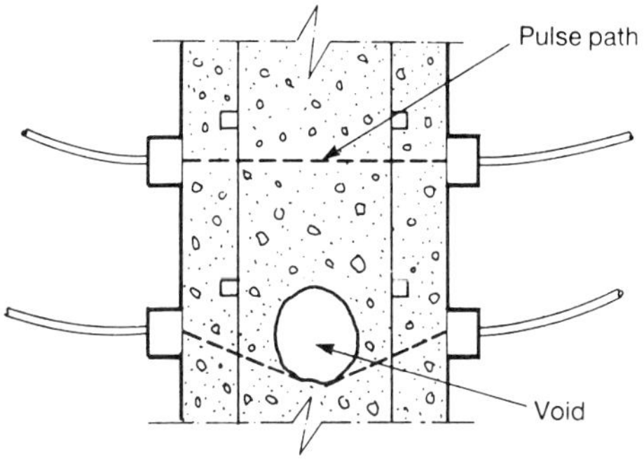

Fig. 3.1. Ultrasonic pulse velocity test (Bungey, 1992)

3.2.3 Ultrasonics

The velocity of an ultrasonic pulse, normally in the range 20 to 50 kHz, is a function of the elastic properties and density of the medium in which it travels. The time taken by an ultrasonic pulse to travel the known distance between the transducers, as illustrated in Figure 3.1, depends on the pulse velocity and the equipment can be calibrated to compare the strength or density at different locations in the concrete or to detect voids or the depth of delamination.

Kay (1992) described fitting the transmitting transducer and receiving transducer to concrete surfaces. The presence of steel reinforcement gives a distorted high value of pulse velocity. However, the results from this technique may be usefully combined with other test data such as that from the Schmidt hammer.

3.2.4 Impact echo

Carino and Sansalone (1990) developed the impact-echo method to detect flaws in concrete. A mechanical impact generates a short duration stress pulse which travels as p- and s-waves which are reflected by discontinuities. A procedure called spectral peak plotting simplifies the interpretation of data from a structure scan. Bungey (1992a) illustrated the method as shown in Figure 3.2.

Krause, Wiggenhauser and Wilsch (1993) presented an advanced pulse echo method for ultrasonic testing of concrete.

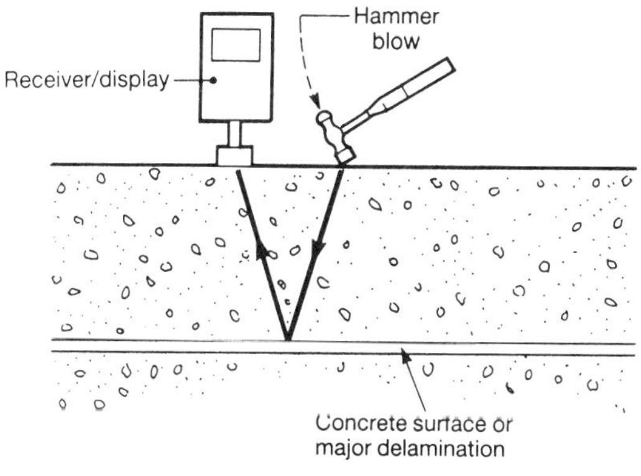

Fig. 3.2. Impact-echo test (Bungey, 1992)

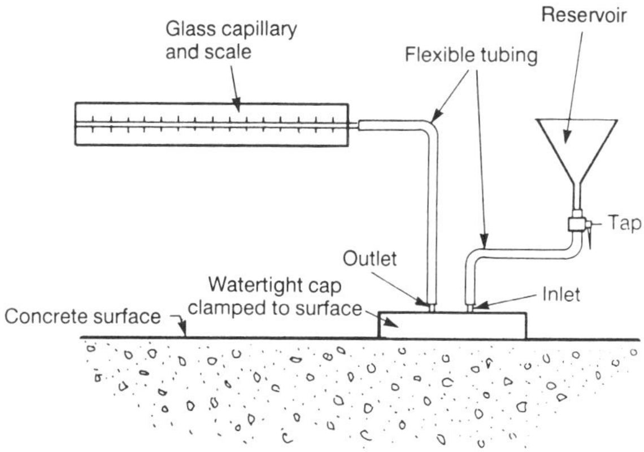

Fig. 3.3. Initial Surface Absorption Test (ISAT) (Bungey, 1992)

3.2.5 Permeability

The Concrete Society (1988) reviewed techniques available for the measurement of concrete permeability. The report covered site and laboratory tests and explained the basic mechanisms and relevant units of measurement. Bungey (1992a) commented that the Initial Surface Absorption Test (ISAT), illustrated in Figure 3.3, is the only method standardized in BS 1881 (1983 and part revision in preparation).

Bungey noted that the Figg method, shown in Figure 3.4, and the Clam method are gaining in popularity.

The moisture content of concrete can be assessed by drilling holes and inserting chemically based disposable meters, electronically based devices or chilled mirror dew-point probes. Leeming (1993) described a surface air permeability test for concrete or surface treatments of concrete.

Fig. 3.4. Figg air test (Bungey, 1992)

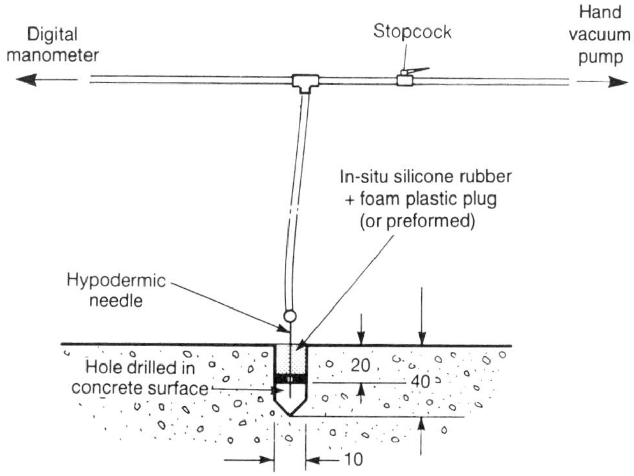

3.2.6 Carbonation

The Building Research Establishment (1981) issued guidance on the phenolphthalein test on freshly broken samples of concrete and (1982) on the diagnosis of corrosion with the help of a flow chart. Bungey (1992a) referred to a more detailed analysis from the microscopic examination of thin sections taken from a small drilled core sample.

3.2.7 Thermography

Clemena and McKeel (1978) used infra-red thermography to define very accurately variations in surface temperatures as a means of detecting areas of delamination due to corrosion in reinforced concrete bridge decks. Warmer distressed areas showed as different colours in a cathode ray tube and were photographed for the record. A comparative study showed the method to be as good as conventional methods of sounding with a hammer or chain drag or a rolling detector but thermography was capable of disclosing areas of incipient delamination.

Temperature differences between sound and unsound concrete may be of the order of 3°C. Although thermography was first used on unsurfaced bridge decks, Manning (1985) has shown that it can also be used on asphalt covered decks, albeit with reduced sensitivity.

3.2.8 Radar

The first use of high frequency electromagnetic waves on concrete has been attributed to Watson (1963) who was measuring moisture content. Clemena (1983) used low power, high resolution ground penetrating radar at 900 MHz to detect delamination in bridge decks. Although not very successful at first, Manning (1985) considered the technique to have good potential, provided the interpretation of wave forms on the oscilloscope could be improved. Whole areas can be scanned by radar at an acceptable speed. Bailey (1989) discussed the theory behind the technique, its advantages and shortcomings. The method is illustrated in Figure 3.5.

Pocock and Hartley (1990) described impulse radar to detect voids, reinforcement, ducts, delaminations or other anomalies. Signal frequency and pulse

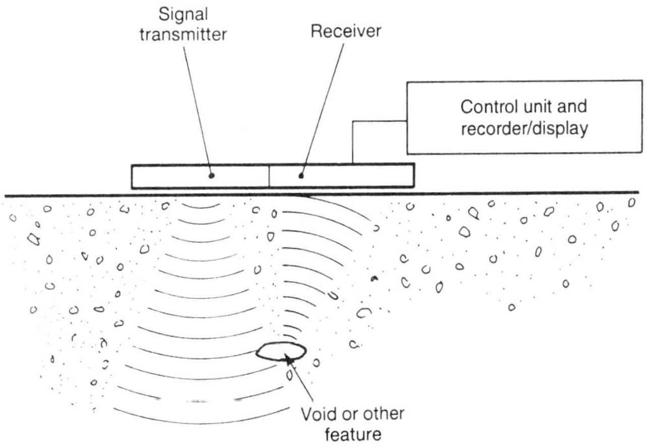

Fig. 3.5. Radar scan (Bungey, 1992)

duration were selected to suit the particular conditions. High resolution could be obtained to depths of 200 mm but accuracy fell off beyond about 500 mm.

Hobbs and Temple (1993) described the following work towards development of the potential of radar in civil engineering applications:

- ☐ signal processing, combined with well-defined inspection procedure, to produce clearer images;
- ☐ modelling to produce theoretical images which can be used to optimize inspections and aid in the interpretation of real data files.

Flohrer and Bernhardt (1993) described the application of radar to detect the location of prestressing tendons behind reinforcement and to detect voids in concrete.

3.2.9 Slot cutting and instrumented coring

The technique developed by the Laboratoire Central des Ponts et Chaussées to measure the existing stress in concrete was reported by Abdunur (1985). A 4 mm wide slot was cut in 10 mm increments to a depth of 80 mm. After each increment a special jack was inserted into the segmental slot and the pressure required to restore the former strain distribution was noted.

Gifford and Partners (1992a) described precision coring a strain gauged area of concrete to estimate the principal stresses present. Elastic constants are obtained by an in-situ jacking test in the core hole. Gifford and Partners (1992b) offered a complementary slot-cutting technique to measure minor transverse principal stresses in narrow prestressed beams. (See also 3.5.)

3.2.10 Other tests

Partly destructive tests of surface strength include pull-out and pull-off tests and penetration tests, described in BS 1881, Part 207 (1992), and by Bungey (1992b) and Kay (1992).

3.3 Reinforcement corrosion

3.3.1 Half-cell potentials

Stratfull (1969 and 1973) reported the significance of chloride in the deterioration of concrete bridge decks and developed the technique of iso-potential mapping. Van Daveer (1975) gave details of the standard copper/copper sulphate half-cell used in North America and the following guidelines to interpret the potential readings:

- more negative than −350 mV, 95% probability of corrosion
- values from −200 to −350 mV, 50% probability of corrosion
- less negative than −200 mV, 5% probability of corrosion.

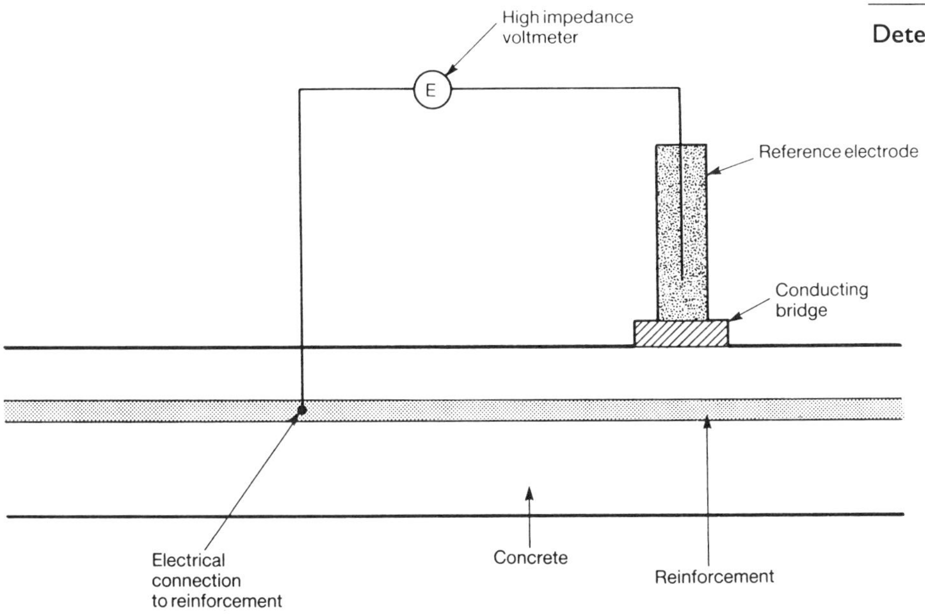

Fig. 3.6. Circuit for measuring half-cell potentials (TRRL Application Guide 9, 1991)

The test procedure, illustrated in Figure 3.6, and other site tests to assess corrosion were described by Vassie (1980 and 1991).

Winnet (1985) developed a microprocessor-controlled multiple half-cell device since which other user-friendly variations of the equipment have become commercially available. One such device, the potential wheel, was described by Langford and Broomfield (1987).

Pocock and Hartley (1990) described the electrochemical mapping technique using a potential wheel and electronic data logging.

Elsener *et al.* (1990) used an eight-wheel electrode system on bridges in Switzerland and found that there are no absolute values of potential to indicate corrosion because the range of potentials indicating corrosion was dependent on factors such as cover, humidity, resistivity, surface layers and chloride concentration. Therefore they preferred to use the structure of the half-cell potential maps and local potential gradients rather than specified critical values of potential.

Naish *et al.* (1990) investigated the effect of moisture on half-cell potentials. Where the surface was dry, the half-cell potential suggested little or no corrosion but when the surface was thoroughly wetted the potential gradients indicated significant macrocell corrosion.

Suryavanshi and Nayak (1990) experimented with a stationary half-cell in place of the usual connection to reinforcement. They found the shape of contours correlated well with those obtained by the normal method although the potential values were completely different and required calibration.

Vassie (1991) confirms the usefulness of Van Daveer's criteria for half-cell potentials in the USA and UK but recognizes the significance of environmental effects, especially moisture. He advises in detail on practical aspects of half-cell potential measurement, interpretation of the data and some supplementary tests to calibrate the values obtained.

3.3.2 Resistivity and the rate of corrosion

Vassie (1980) noted that, as corrosion is an electrochemical process, the probability of corrosion is proportional to the corrosion current which is controlled by the resistivity of the concrete. He described the measurement of the resistivity of concrete by an adaptation of a method proposed by Wenner (1915) for soils.

Four equidistant contact points are required in a straight line on the concrete surface. He recommended holes 6 mm deep to take 6 mm diameter electrodes. Resistivity is calculated from the potential between the inner electrodes when an alternating current is passed between the outer electrodes. The method is illustrated in Figure 3.7.

The Department of Industry (1981) outlined four methods of detecting and monitoring corrosion:

- ☐ half-cell potentials
- ☐ constant anodic current polarization
- ☐ electrical resistance probe
- ☐ polarization resistance probe.

All methods were applicable in situ but methods (*b*) and (*c*) required electrodes to be cast into the concrete and method (*d*) was subject to the resistivity being no higher than 10 000 Ωcm which could be a problem in dry concrete.

Escalante and Ito (1981) measured corrosion rates using polarization techniques. They found that corrosion of steel in chloride-free concrete was less

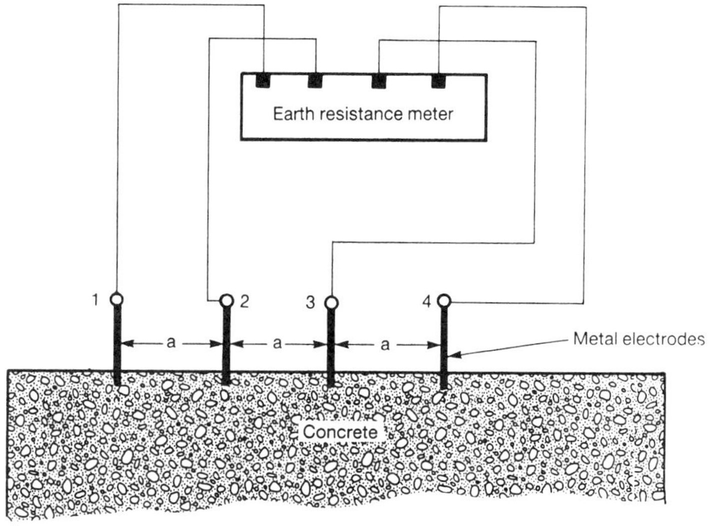

Fig. 3.7. Wenner four electrode resistivity measurement (TRRL Laboratory Report 953, 1980)

than 0.007 mg/dm^2d but in chloride-contaminated concrete the average rate was an order of magnitude greater at 0.33 mg/dm^2d.

Tilly (1990) related the rate of corrosion to its likelihood as given by Vassie (1980) and summarized in Table 3.1.

Table 3.1. Resistivity as an indicator of corrosion

Resistivity (Ω cm)	Likelihood	Rate of corrosion
Greater than 12 000	Unlikely	Low
5 000 to 12 000	Probable	Intermediate
Less than 5 000	Almost certain	High

Pocock and Hartley (1990) described polarization resistance measurements and the galvanostatic pulse method which was being developed to measure rates of corrosion but noted that there was no equipment available for use on site.

Vassie (1991) described electrode potentials and current flow, anodic and cathodic current densities and the ratio of the latter, known as the pitting ratio.

3.3.3 Chloride concentrations

Clear (1974) emphasized the need to identify contaminated concrete and determine chloride concentrations. He discussed corrosion thresholds and the removal of concrete and presented the wet chemical analysis procedure of H. A. Berman applied to samples obtained with a rotary hammer.

The Building Research Establishment (1982) described some effects of chloride contamination combined with carbonation.

Vassie (1980) described the Quantab method recommended for site determination of total chloride content and a variation to find the free chloride content by careful extraction with water instead of acid.

Figg (1986) reviewed available methods and emphasized the importance of selecting 40 samples from representative locations for one structure and considered that present operation times could be greatly reduced by improvements in techniques in the sampling and analysis.

Page (1987) reviewed several areas of current research related to chloride-induced corrosion, non-destructive electrochemical techniques for monitoring corrosion rates and chloride-induced pitting with reference to composition of the pore solution.

Konda (1990) described:

- ☐ sampling and sample preparation of hardened concrete for chloride analysis
- ☐ implications of total chloride and free chloride limits in specifications.

Kay (1992) offered guidance on drill dust sampling and typical methods are shown in Figures 3.8 and 3.9.

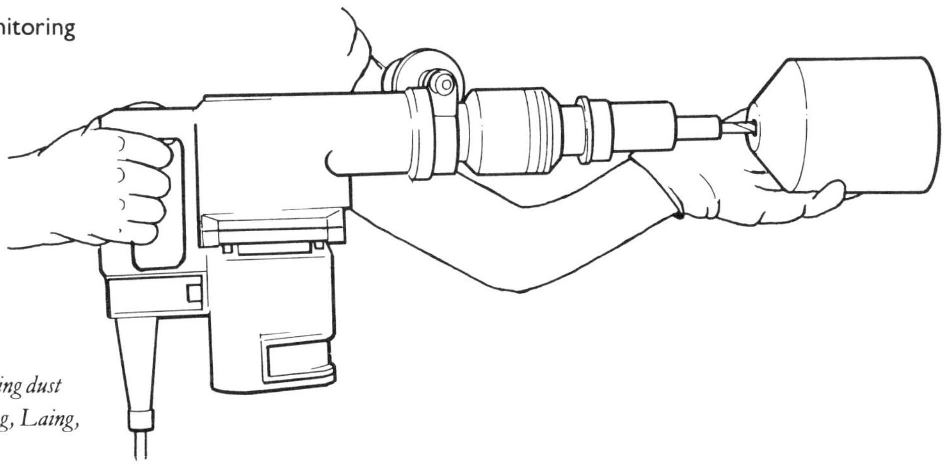

Fig. 3.8. Sampling drilling dust with plastic shroud (Figg, Laing, 1986, TRRL CR 32)

3.3.4 Interpretation of results

No single type of test is sufficient to determine the state of corrosion. Some combination will usually be required of:

(*a*) visual assessment (3.1.2)
(*b*) depth of cover (3.2.1)
(*c*) half-cell potentials (3.3.1)
(*d*) resistivity (3.3.2)
(*e*) chloride concentrations (3.3.3) and, possibly,
(*f*) carbonation (3.2.6).

Clear (1974) regarded methods (*a*)–(*c*) and (*e*) above, together with the delamination detector, as the essential tools in the evaluation process. He emphasized that the critical chloride concentration was that which destroys the passive con-

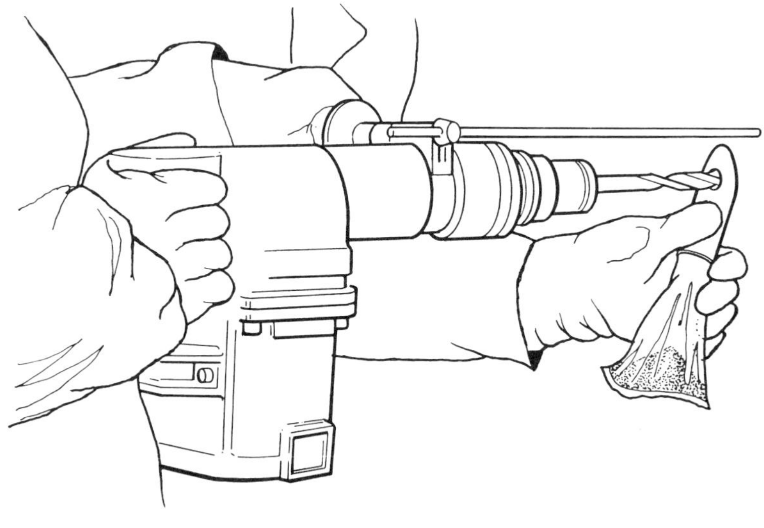

Fig. 3.9. Sampling drilling dust with pierced plastic tube (Figg, Taywood, 1986, TRRL CR 32)

crete environment around the reinforcement. In other words, it is more important to prevent the initiation of corrosion than to study the rate of corrosion.

Brown, Geoghegan and Baker (1983) reviewed testing techniques and some effects which may lead to misinterpretation of the data. For example, chloride contamination or carbonation may distort surface potentials as much as 100 mV. They considered simple damage theory with t_0 the time to initiation of corrosion, and t_1 the time for damage to become in need of repair.

Skovesende (1989) described a device for logging both half-cell potential and resistivity simultaneously to produce contour diagrams. He reported good correlation between half-cell potentials, resistivity, moisture and chloride contents. He considered resistivity to be an important check on the reliability of half-cell potential measurements.

3.3.5 Monitoring reinforced concrete

Vassie (1980) described resistance probes and linear polarization probes to measure rates of corrosion but there are limitations, not least the fact that probes must be installed during construction.

Lambert and Page (1986) described three electrochemical monitoring techniques to make rapid assessments of the condition of reinforcement in large structures which have not deteriorated to the stage where corrosion is evident. The three methods are:

- half-cell potential mapping
- AC and DC polarization of the steel
- electrochemical noise.

Langford and Broomfield (1987) described some equipment and techniques available for monitoring corrosion in reinforced concrete, indicating how linear polarization, Corrosimeter, Sereda and Beeby corrosion probes can be integrated to form a corrosion monitoring station.

John *et al.* (1987) described a 'Guard Ring' method for the inspection or monitoring of reinforcement corrosion. A central auxiliary electrode polarizes the reinforcement locally while the outer ring simultaneously supplies polarization to the surrounding mat. This was used successfully for linear polarization and electrochemical impedance determinations. Absolute polarization resistance cannot normally be obtained on steel in concrete but apparent polarization resistance can be used to assess the comparative corrosion rates.

Elsener *et al.* (1988 and 90) reported potential mapping of Swiss bridge decks using a combination of local potential gradient with resistivity to give a rough estimate of the corrosion rate in a single macrocell. They also reported a very rapid technique, galvanostatic pulse measurement, giving unambiguous results when half-cell potential measurements are uncertain.

Smith (1988) reviewed the literature on detecting and measuring corrosion of reinforcement in concrete. The methods for estimating corrosion rates were given as follows:

- measurement of potential and potential mapping

- electrochemical noise
- electrical resistance
- constant anodic polarization
- linear polarization resistance
- embedded probe.

However, he considered no one of these to be reliable.

In a programme of monitoring selected crossheads in the Midland Links viaducts, half-cell potential surveys and chloride concentration data were collected at regular intervals and fed into a database. As the contour maps from multiple surveys were unwieldy to compare, the consultants, in conjunction with TRRL, developed a computer program described by Gower and Calder (1989). The program CONTEST handles data as cumulative frequency distributions (S curves) of electro-potential and chloride concentrations as histograms.

The following applications of CONTEST were suggested.

- Determination of repair criteria and the effect of different criteria on the volume of concrete to be removed.
- Assessment of the effectiveness of repairs and the investigation of the formation of secondary anodes.
- Assessment of seasonal variations, optimum frequency for monitoring and its consequences for repair criteria.
- Assessment of long-term changes in potentials after drainage improvements.
- Revision of condition classifications.
- Analysis of half-cell potential data and chloride concentrations in relation to corrosion and pitting measured after concrete removal.
- Evaluation of half-cell potentials after application of various surface treatments.

Pocock and Hartley (1990) described a nitrogen gas pressure test on standard cores which could be carried out on site to detect ASR before surface cracking is evident.

Wolff and Miesseler (1990 and 1993) described two methods of monitoring concrete structures over a long time:

- optical fibre sensors
- copper wire sensors.

Nowac (1990) emphasized the need for efficient diagnostic procedures to evaluate the load capacity of a structure, to monitor its load and resistance history and to evaluate accumulated damage. He reviewed the techniques available and considered signal processing, failure detection and isolation and current trends in the development of new techniques.

Pichot and Trouillet (1990) presented a non-destructive imaging technique using microwaves in the X-band region from 7 to 13 GHz. Its performance is compared with that of magnetic metal detectors.

Grimaldi *et al.* (1990) developed a set of methodologies and facilities to determine the areas of a bridge to be repaired using measurements of permeability, carbonation and half-cell potentials of reinforcing bars close to one another.

Dawson, John *et al.* (1990) presented electrochemical techniques to monitor reinforcement corrosion on site. Electrochemical noise can indicate pitting and linear polarization resistance measurement or current mapping can indicate the rate of corrosion.

Wood (1990) considered that the only reliable way to predict the ingress of carbonation and/or chloride contamination is from service performance. Rather than coarse grid overall sampling, he proposed that carefully selected representative sample areas should be tested, measuring carbonation depths and/or chloride profiles together with a covermeter histogram for each area sampled. Once results have been obtained in this way for several years, the rate of carbonation and/or ingress of chloride can be calculated and the time to initiation of corrosion can be predicted with some confidence. As the rate of corrosion thereafter is less definite, initiation should be taken to be the durability limit state.

Schiessl and Raupach (1993) reported on permanent monitoring of the corrosion risk for reinforcement.

Wiggenhauser *et al.* (1993) presented a method for on-line analysis of the distribution of chemical elements on the surface. The technique employs time-resolved fluorescence spectroscopy after the vaporization of minute amounts of material by very intense laser shots scanned over the surface. Elements of special interest are those linked with aggressive agents, such as sodium, linked with deicing salt, and it is claimed that two-dimensional distributions on the surface can be measured and documented.

3.4 Post-tensioning tendons

3.4.1 Exploratory hole drilling

Woodward (1981) described conditions within post-tensioning ducts using the exploratory hole drilling technique. Tilly (1990) rated this the most direct and practical method of inspecting post-tensioning ducts. Holes are drilled carefully into the duct and, if voids are present, these are inspected and photographs of the tendons can be taken. Price (1986) described the investigation of bridges in Oxfordshire, drilling holes up to 200 mm deep with a success rate better than 75% in reaching targets set out with measurements from drawings. To estimate the size of the void the vacuum/pressure technique can be conducted as follows.

(*a*) Up to five holes are drilled into the top of each duct.
(*b*) A vacuum is applied to each hole in turn and pressures checked at the others to test the continuity of voids.
(*c*) The volume of an evacuated void is measured by means of a connecting perspex tube dipping into water. The height to which the water is drawn up the tube can be used to calculate the volume of the void.

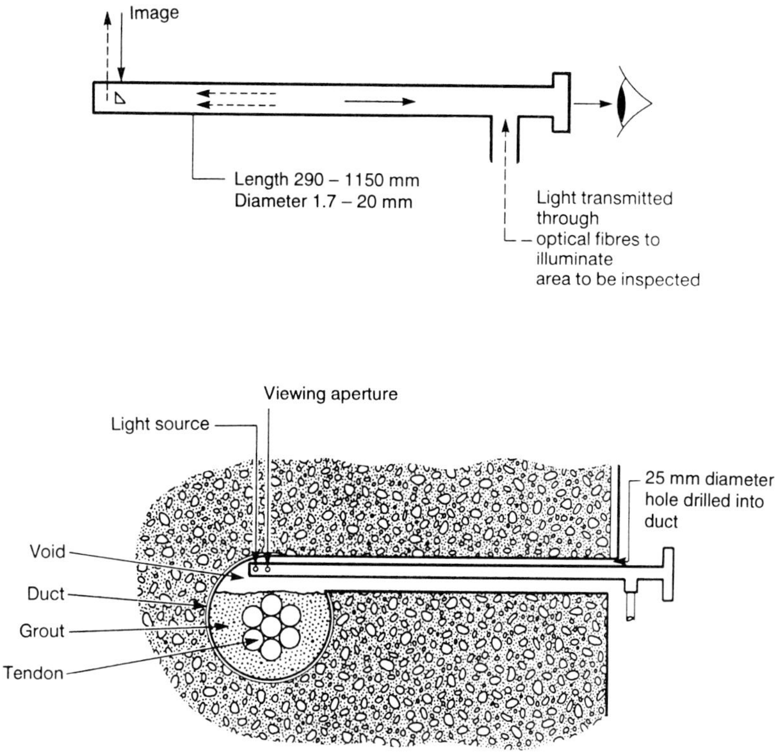

Fig. 3.10. Endoscope inspection (Woodward, 1981)

(*d*) Air leakage from a duct to the atmosphere is measured by pressurizing the duct and measuring the flow rate.

(*e*) Where a void is large enough, the field of vision may be illuminated by means of fibre optics and viewed and photographed through an endoscope as in Figure 3.10.

The method only detects broken wires if they are present at the point inspected.

New Civil Engineer (1993) reported a combination of impulse radar (3.2.8) and hole drilling to locate post-tensioning ducts. The small diameter percussion drill used was linked to a visual alarm to prevent damage to the tendon.

3.4.2 Radiography

The loss of energy, or attenuation, of gamma rays passing through a heterogeneous medium is greater in parts of higher density. The source, usually cobalt 60 or iridium 192, is set up on one side of the concrete to be examined and a suitable photographic film on the other side as shown in Figure 3.11. The thickness of concrete penetrated is limited to 500 mm.

High energy X-rays have been used successfully for thicknesses of concrete greater than one metre by Zouboff (1973) and by Pullen and Clayton (1981). X-rays can be switched on and off.

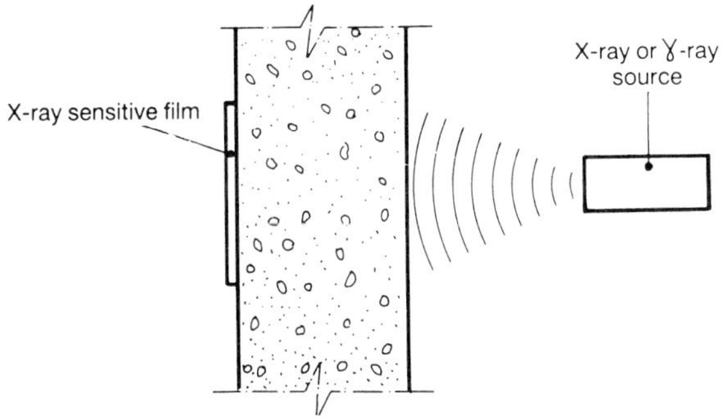

Fig. 3.11. Radiography scan (Bungey, 1992)

Guinez, Chatelain and Chevrier (1989) outlined how the quality of grouting in ducts had been verified in France since 1968 by gammagraphy and since 1985 by radioscopy using a miniaturized linear accelerator.

Radiography may be useful to give a picture of a fairly small area, showing the position and size of bars and tendons, but it will not show corrosion or broken wires and the definition may be poor. The method is expensive and time-consuming, the photographs may be difficult to interpret and strict safety precautions must be observed.

3.4.3 Ultrasonics

If the anchorages are accessible, an alternative approach is to propagate high frequency pulses along the tendon. Initial attempts under contract to TRRL could only detect fractured wires within a few metres but the contractor, the Electrical Research Association (1984), suggested some ways in which the technique could be improved to detect fractures in tendons up to 20 m in length. However, even if the technique were to be improved, the ends of tendons are generally inaccessible as the usual practice was to encase them with in-situ concrete.

3.4.4 Monitoring tendons

The Standing Committee on Structural Safety (1979–89 and 1992) has examined the safety of bridges having grouted post-tensioned steel tendons, noting the particular technical difficulties of determining their structural condition. SCOSS (1989) had to conclude that

> 'no suitable NDT is yet available and none is likely to be developed in the near future. However ... a sufficient body of experience exists to enable guidelines to be drawn up for structural appraisal, including the use of drilling to investigate tendon condition'.

Woodward and Wilson (1991) described load tests on edge beams taken from the collapsed Ynys-y-Gwas bridge and a supporting analytical study. They reached the following conclusions.

(*a*) Corrosion of tendons can cause sudden collapse of precast prestressed segmental bridges.
(*b*) Localized corrosion that has not yet caused failure of a wire or tendon is undetectable by monitoring deformation or joint opening.
(*c*) It is only where breakages occur without collapse of the structure that they might be detected by monitoring deformation. This would give no indication of the degree of corrosion and such a structure should be considered to be on the point of collapse.
(*d*) Progressive failure of one edge beam was due to successive failure of wires or tendons. This ductile mode cannot be relied on unless the structure can shed load so that sudden collapse does not occur.
(*e*) Tests to failure of the two edge beams were simulated by mathematical models. Shapes of moment-deflection curves were governed by

(i) linear and non-linear elastic properties
(ii) prestressing force
(iii) bond between tendons, grout and concrete
(iv) partial failure of prestress before collapse.

(*f*) The degree of corrosion at joints did not alter the shape of the curves but did influence the failure load.
(*g*) The prestressing force governs the onset of non-linear behaviour. Before this, behaviour is independent of the degree of corrosion, regardless of whether or not fractures have occurred.
(*h*) Bond strength was irrelevant to the detection of corrosion, except that weak bond might help to detect failures. Load testing beyond the elastic limit would cause irreparable damage to bond between tendons and concrete and is positively dangerous.
(*i*) Fractures of wires or tendons might be detected by monitoring very accurately the 'at rest' position of a structure.

SCOSS (1992) recommended that a stringent regime of inspection and appraisal of existing bridges be introduced, particular care being taken to determine

> 'the implications of tendon failure due to corrosion. Where the implications were severe, additional structural capacity to offset the potential structural weakness should be installed'.

Wolff (1990) described optical fibre sensors and copper wire sensors applied to the long-term monitoring of 'intelligent' glass fibre prestressing tendons.

Darby (1992) reflected the heightened concern of engineers, since the Ynys-y-Gwas bridge collapse in 1985, to safeguard any post-tensioned bridges for which they are responsible. Three further cases had come to light in Oxfordshire. The drilling and vacuum/pressure techniques developed since the 1970s had yielded much useful information but voids in the ducts and corrosion of the tendons did not necessarily go together. Defects in tendons could be very localized and anchorages are the most vulnerable and the most difficult locations to inspect. He therefore recommended the following action.

- Assess the factors influencing corrosion and the sensitivity of the structure to loss of prestress.
- Inspect vulnerable areas, including drilling and contamination sampling.
- If prestress may have been lost, assess remaining prestress by coring or slotting of concrete (3.2.9, 3.5) or by drilling of reinforcement (3.5).
- If future durability is in doubt, install a monitoring system such as an optical fibre sensor technique.

3.5 In-situ stress determination

Owens (1993) described the techniques developed by Strainstall Engineering Services Ltd and an assessment of their accuracy in a programme of work carried out in conjunction with TRL. The centre hole technique presented by Beaney and Proctor (1974) depends upon the stress relief caused by drilling a relatively small hole in a uniformly stressed material. The radial stress at the edge of the hole is zero—total relief of stress—and the relief of stress falls off as the distance from the hole increases, typically as shown in Figure 3.12.

From measurements of the strain change at a distance of about two hole diameters from the centre of the hole it is possible to deduce the uniform stress prior to drilling.

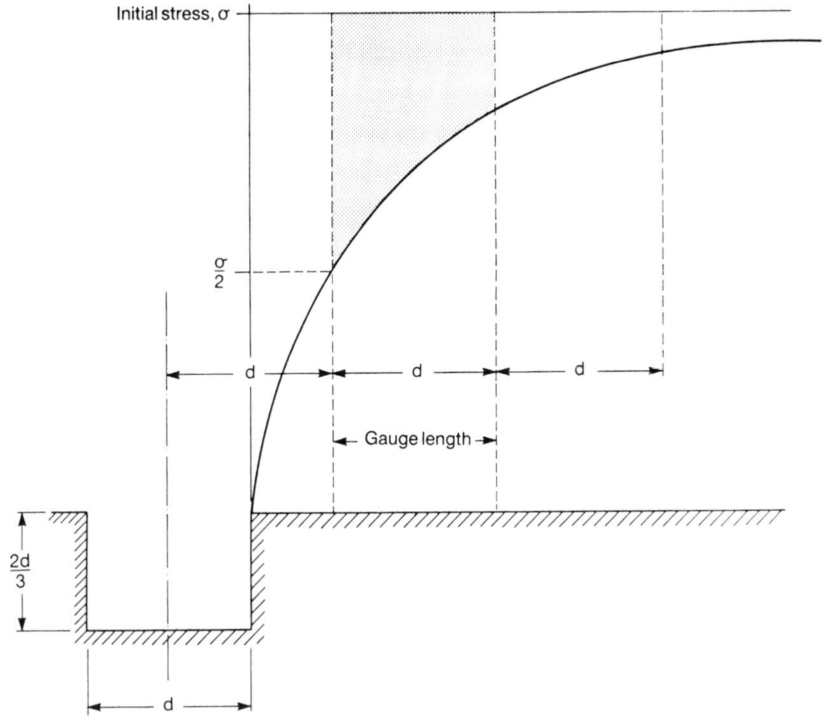

Fig. 3.12. Stress relief caused by drilling (Owens, 1993, Strainstall Engineering Services)

The technique can be applied to concrete, provided the location is clear of reinforcement. The hole size should be at least twice the aggregate size and diamond drilling a 67 mm diameter core leaving a 76 mm diameter hole has been found to be effective. 50 mm vibrating wire gauges are attached to the concrete surface by adhesive at the centre of and radially outside the hole location. Incremental readings are taken as the hole is drilled to a depth of 50 mm, the core is removed, the bottom of the hole flattened to receive another gauge and further measurements taken as the drilling is continued a further 50 mm. The elastic modulus is obtained by testing the core. Accuracy of principal stress measurement is ± 0.3 N/mm^2 for elastic modulus 34 N/mm^2 or $\pm 3\%$, whichever is greater.

To determine the existing stresses in steel reinforcement or prestressing wires, 1.6 mm diameter holes can be drilled to a depth of 1 mm and the strains measured with miniaturized electrical resistance foil strain gauges. Accuracy in location and drilling are paramount. A carbide burr in a very high speed air turbine induces no significant drilling strain. The accuracy of the stress determination is ± 7 N/mm^2 for elastic modulus 207 kN/mm^2 or $\pm 3\%$, whichever is greater. Typically, locked in manufacturing stresses are 15 ± 7 N/mm^2 for reinforcement and 30 ± 7 N/mm^2 for prestressing wires. The overall accuracy for determining the effective stress in the steel is ± 14 N/mm^2 or $\pm 4\%$, whichever is greater.

Drilling the small hole is semi-destructive but the reduction in cross-section is only 8% for a 5 mm diameter wire or 4% for a 7 mm diameter wire. This degree of loss of strength in one or two wires of a seven-wire strand or a twelve-wire tendon is not critical.

3.6 References

ABDUNUR C. (1985). *Stress measurement in structures by a method of miniaturized stress release*. Bulletin de Liaison LCPC, No. 138, July–Aug., 5–14.

ADDERSON B. W. (1986). *A short literature survey of the application of high frequency electromagnetic waves in the non-destructive testing of materials*. BRE Note N86/86.

AGAR T. J. A. and KENDALL C. J. (1985). Computer storage, management and analysis of data arising from investigations into structural deterioration. *Proc. 2nd Int. Conf. on Civil and Structural Engineering Computing*. **1**, 3–5 Dec., Institution of Civil Engineers, 131–134.

ALLDRED J. C. (1993). An improved method for measuring reinforcing bars of unknown diameter using a covermeter. *Proc. 2nd Int. Conf. on Bridge Management*. University of Surrey, Guildford, 18–21 Apr.

BAILEY M. (1989). The use of ground penetrating impulse radar in the investigation of structural faults. *Proc. Int. Conf. on Structural Faults and Repair*. Kensington, London, 27–29 June, 11–18.

BARRY J. (1989). Access to access. *Proc. Int. Conf. on Structural Faults and Repair*. Kensington, London, 27–29 June, 31–32.

BEANEY E. M. and PROCTOR E. (1974). A critical evaluation of the centre hole technique for the measurement of residual stress. *Strain*. Jan.

BEEBY A. (1978). Corrosion of reinforcing steel and its relation to cracking. *Structural Engineer*, **56A**, Mar., 77–81.

BICKLEY J. A., EMERY J. J., STURRUP V. and MACDONALD P. (1986). *Repair techniques for concrete bridge components.* RTAC Report 1029, Roads and Transportation Association of Canada.

BORGARD B., WARREN C., SOMAYAJI S. and HEIDERSBACH R. (1989). *Correlation between corrosion of reinforcing steel and voids and cracks in concrete structures.* Transportation Research Record 1211.

BOYES D. S. (1990). On which bridges should we spend our money? *New concepts for the management of highway structures.* Institution of Highways and Transportation National Workshop. 20 Mar., Leamington Spa.

BRIDGE INSPECTION PANEL (1984). *Bridge inspection guide.* Department of Transport, Scottish Development Department, Welsh Office and Department of the Environment for Northern Ireland. HMSO, London.

BRITISH STANDARDS INSTITUTION (1986). *BS 1881 Testing concrete.* Part 201. Guide to use of NDT methods for hardened concrete. Part 203. Ultrasonic Pulse Velocity. Part 205. Radiography. BSI, London.

BRITISH STANDARDS INSTITUTION (1988). *BS 1881 Testing concrete.* Part 204. Recommendations on the use of electromagnetic covermeters, BSI, London.

BRITISH STANDARDS INSTITUTION (1992). *BS 1881 Testing concrete.* Part 207. Near to surface strength tests. Part 208. Initial surface absorption (revision of Part 5). BSI, London.

BROWNE R. D., GEOGHEGAN M. P. and BAKER A. F. (1983). Analysis of structural condition from durability results. Corrosion of reinforcement in concrete construction. A. P. Crane (ed.). Meeting 13–15 June, London. Published for Society of Chemical Industry by Ellis Horwood Ltd, Chichester, 193–222.

BUILDING RESEARCH ESTABLISHMENT (1981). *Carbonation of concrete made with dense natural aggregates.* Roberts M. H. Information paper 6/81.

BUILDING RESEARCH ESTABLISHMENT (1982). Part 2. The durability of steel in concrete: *Diagnosis and assessment of corrosion-cracked concrete.* Digest 264.

BUNGEY J. H. (1992a). Developments in testing of concrete in structures. *Symp. on the Durability of Highway Structures.* Institution of Highways and Transportation, Taunton, 12 Nov.

BUNGEY J. H. (1992b). *Testing concrete in structures. A guide to equipment for testing concrete in structures.* CIRIA Technical Note 143, 87.

CALDWELL J. A. D. and SCALES R. (1989). Recently developed methods for the assessment of concrete structures. *Proc. 3rd Int. Conf. on Deterioration and Repair of Reinforced Concrete in the Arabian Gulf.* Bahrain Society of Engineers, 21–24 Oct., **1**, 411–425.

CARINO N. J. and SANSALONE M. (1990). A. S. Nowak (ed.). Flaw detection in concrete using the impact-echo method. *Bridge evaluation, repair and rehabilitation.* Nato ASI Series E: Applied Sciences, Vol. 187, ISBN 0 7923 0999 5, 101–118.

CEMENT and CONCRETE ASSOCIATION (now BRITISH CEMENT ASSOCIATION) (1988). Diagnosis of alkali-silica reaction.

CLEAR K. C. (1974). Evaluation of Portland cement concrete for permanent bridge deck repair. FHWA-RD-74-5. Federal Highway Administration, Washington DC, USA.

CLEMENA G. G. and MCKEEL W. T. jr (1978). *Detection of delamination in bridge decks with infrared thermography.* Transportation Research Record No. 664, Vol. 1, 180–182.

CLEMENA G. G. (1983). *Non-destructive inspection of overlaid bridge decks with ground penetrating radar.* Transportation Research Record No. 899, 21–34.

CONCRETE SOCIETY (1987). *Alkali-silica reaction, minimising the risk of damage to concrete.* TR.030. ISBN 0 946691 20 7.

CONCRETE SOCIETY (1988). *Permeability testing of site concrete—review of methods and experience.* TR.031. ISBN 946691 21 5.

CONCRETE SOCIETY (1992). *Non-structural cracks in concrete*. TR.022. ISBN 0 946691 43 6.

COPE R. J. and BUNGEY J. H. (1987). Assessment of concrete in bridge structures. *Concrete bridge engineering: performance and advances*. Elsevier Applied Science, 25–62.

DARBY J. J. (1992). Maintenance—a county viewpoint. *Symp. on the Durability of Highway Structures*. Institution of Highways and Transportation, Taunton, 12 Nov.

DARWIN D., MANNING D., HOGNESTAD E., BEEBY A., RICE P. and GHOWRWAL A. (1985). Debate: crack width, cover and corrosion. *Concrete International*, May, 20–35.

DAWSON J. L., JOHN D. G., JAFAR M. I. and SHERWOOD L. (1990). Electrochemical methods for the inspection and monitoring of corrosion of reinforcing steel in concrete. *Proc. 3rd Int. Conf. on Corrosion of Reinforcement in Concrete Construction*. Wishaw, Warwickshire, 21–24 May. Published for the Society of Chemical Industry by Elsevier Applied Science, 358–371.

DEPARTMENT of INDUSTRY (1981). *The corrosion of steel, and its monitoring, in concrete*. Guide to practice in corrosion control No. 7.

DEPARTMENT of TRANSPORT (1977). *The inspection of highway structures*. Technical Memorandum (Bridges), BE4/77.

DEPARTMENT of TRANSPORT (1986). *The investigation and repair of concrete highway structures*. Departmental Advice Note BA23/86.

DEPARTMENT of TRANSPORT (1988). *Trunk road and motorway structures—records and inspection*. Trunk Road Maintenance and Management Notice TRMM 2/88.

ELSENER B., MÜLLER S., SUTER M. and BÖHNI H. (1988). Potential mapping of steel in concrete. *Proc. Int. Conf. of the Institution of Corrosion Science and Technology and the National Association of Corrosion Engineers*. Brighton, 3, 4 Oct., **3**, 169–181.

ELSENER B., MÜLLER S., SUTER M. and BÖHNI H. (1990). Corrosion monitoring of steel in concrete—theory and practice. *Proc. 3rd Int. Conf. on Corrosion of Reinforcement in Concrete Construction*. Wishaw, Warwickshire, 21–24 May. Published for the Society of Chemical Industry by Elsevier Applied Science, 348–357.

ESCALANTE E. and ITO S. (1981). Measuring the rate of corrosion of steel in concrete. *Corrosion/81*. National Association of Corrosion Engineers. 6–10 Apr., Sheraton Centre, Toronto.

FIGG J. W. and MARSDEN A. F. (1985). *Development of inspection techniques for reinforced concrete: a state of the art survey of electrical potential and resistivity measurements for use above water level*. Concrete in the Oceans Technical Report (OTH) 10 84 205. HMSO, London.

FIGG J. W. (1986). *Methods to determine chloride concentrations in in situ concrete*. TRRL Contractor Report 32.

FLOHRER C. and BERNHARDT B. (1993). Detection of prestressed steel tendons behind reinforcing bars, detection of voids in concrete structures—a suitable application for radar systems. *Proc. 2nd Int. Conf. on Bridge Management*. University of Surrey, Guildford, 18–21 Apr.

FORDE M. C. (ed.) (1989). *Proc. Int. Conf. on Structural Faults and Repair*. Kensington, 27–29 June.

FREARSON J. P. H. (1985). *The problem—identification and interpretation*. Proc. of a one-day Symposium. London Tara Hotel, Concrete Society, 25 Sept.

GIFFORD AND PARTNERS (1992a). Instrumented coring. Capability profile.

GIFFORD AND PARTNERS (1992b). Slot cutting for measurements of concrete stresses. Capability profile.

GOWER M. R. and CALDER A. J. J. (1989). Monitoring and evaluating reinforcement corrosion using CONTEST. *Proc. 3rd Int. Conf. on Deterioration and Repair of Reinforced Concrete in the Arabian Gulf*. Bahrain Society of Engineers.

GRIMALDI G., BERISSI R., BREVET P. and RAHARINAIVO A. (1990). A. S. Nowak (ed.).

Techniques for assessing the concrete (state) condition in bridges. *Bridge evaluation, repair and rehabilitation*. Nato ASI Series E: Applied Sciences, Vol. 187, ISBN 0-7923-0999-5, 249-258.

GUINEZ R., CHATELAIN J. and CHEVRIER J-P. (1989). Radioscopy of prestressed concrete bridges. *Symp. on the Durability of Structures*. Lisbon. International Association for Bridge and Structural Engineering, 347-352.

HARDING J. E. and PARKE G. A. R. (eds) (1990). *Proc. 1st Int. Conf. on Bridge Management*. University of Surrey, Guildford, 28-30 Mar., Elsevier Applied Science. ISBN 1 85166 456 4.

HARTT W. H., DUNN S. E., WENG M. S. and BROWN R. P. (1981). *Applications of acoustic emission to detection of reinforcing steel corrosion in concrete*. Paper No. 49 presented to the International Corrosion Forum, 6-10 Apr., Toronto.

HOBBS C. P. and TEMPLE J. A. G. (1993). Radar inspection of civil engineering structures. *Proc. Int. Conf. on NDT in Civil Engineering*. University of Liverpool, 14-16 Apr.

JOHN D. G., EDEN D. A., DAWSON J. L. and LANGFORD P. E. (1987). Corrosion measurements on reinforcing steel and monitoring of concrete structures. *Proc. Symp. on Corrosion of Metals in Concrete*. National Association of Corrosion Engineers, Texas, 159-167.

KAY T. (1992). *Assessment and renovation of concrete structures*. Longman Scientific and Technical, Harlow.

KENT D. (1985). Integrated tests ensure good maintenance repair. *Surveyor*, 20 June, **165**, No. 4849, 15-16.

KONDA M. V. V. S. (1990). Chloride in concrete. (a) Sampling of hardened concrete for chloride analysis. (b) Limits on chloride content in reinforced concrete and their meaning in practical terms. *New Zealand Concrete Construction*. Dec., 14-17.

KRAUSE M., WIGGENHAUSER H. and WILSCH G. (1993). Advanced pulse echo method for ultrasonic testing of concrete. *Proc. Int. Conf. on NDT in Civil Engineering*. University of Liverpool, 14-16 Apr.

LAMBERT P. and PAGE C. L. (1986). The monitoring of reinforcement corrosion. *Proc. of a Research Seminar on the Behaviour of Concrete Structures*, 30 June-2 July. Cement and Concrete Association, 288-291.

LANGFORD P. and BROOMFIELD J. (1987). Monitoring the corrosion of reinforcing steel. *Construction Repair*, May, **1**, No. 2, 32-36.

LEEMING M. B. (1993). A surface air permeability for concrete and surface treatments on concrete. *Proc. 2nd Int. Conf. on Bridge Management*. University of Surrey, Guildford, 18-21 Apr.

MCGRATH B. J. and SCHUTT W. R. (1989). Testing and analysis of corrosion in steel reinforced concrete structures. *Proc. 3rd Int. Conf. on Deterioration and Repair of Reinforced Concrete in the Arabian Gulf*. Bahrain Society of Engineers, 21-24 Oct., **1**, 401-410.

MALLETT G. P. (1986). Bridge maintenance—value for money. Institution of Highways and Transportation National Workshop. *Bridge maintenance initiatives*. Leamington Spa, 8 Apr.

MANNING D. G. (1985). *Detecting defects and deterioration in highway structures*. NCHRP synthesis of highway practice 118. Transportation Research Board, Washington, USA.

MARKS V. J. (1977). *Detection of steel corrosion in bridge decks and reinforced concrete pavements*. Report No. 156, Iowa Department of Transportation, USA.

MAYS G. (ed.) and BAKER A. F. (1992). Structural investigations. *Durability of concrete structures: investigation, repair and protection*. E. and F. N. Spon, 37-81.

MOORE W. M. (1973). *Detection of bridge deck deterioration*. Highway Research Record No. 451, Washington, USA, 53-61.

NAISH C. C., HARKER A. and CARNEY R. F. K. (1990). Concrete inspection: interpretation of potential and resistivity measurements. *Proc. 3rd Int. Conf. on Corrosion of Reinforcement in Concrete Construction.* Wishaw, Warwickshire, 21–24 May. Published for the Society of Chemical Industry by Elsevier Applied Science.

NEVILLE A. (1981). *Properties of concrete.* Pitman, London, 301.

NOWAC A. S. (ed.) (1990). *Proc. of NATO Advanced Research Workshop on Bridge Evaluation, Repair and Rehabilitation.* Baltimore, Maryland, USA. 30 Apr.–2 May. ISBN 0 7923 0999 5. Kluwer Academic Publishers, Dordrecht, 3300 AA, Netherlands.

NOWAC A. S. (1990). A. S. Nowak (ed.). Diagnostic procedures for bridges. *Bridge evaluation, repair and rehabilitation.* Nato ASI Series E: Applied Sciences. Vol. 187. ISBN 0 7923 0999 5, 73–84.

OECD (1981). *Bridge maintenance.* Road Research Group Report. Organisation for Economic Co-operation and Development, 2, rue André Pascal, 75775 Paris Cedex 16.

OECD (1992). *Bridge management.* Scientific Expert Group Report. Organisation for Economic Co-operation and Development, 2, rue André Pascal, 75775 Paris Cedex 16.

OWENS A. (1993). In situ stress determination used in structural assessment of concrete structures. *Annual Conference of the British Society of Strain Measurement.* Oxford, Sept.

PAGE C. L. (1987). An overview of current research. *Proc. 2nd Int. Conf. on Deterioration and Repair of Reinforced Concrete in the Arabian Gulf.* Bahrain Society of Engineers. 11–13 Oct, Vol. 2, 161–171.

PICHOT C. and TROUILLET P. (1990). A. S. Nowak (ed.). Diagnosis of reinforced structures: an active microwave imaging system. *Bridge evaluation, repair and rehabilitation.* Nato ASI Series E: Applied Sciences. Vol. 187, ISBN 0 7923 0999 5, 201–215.

POCOCK D. C. and GIBBS B. (1986). Diagnosis, prevention and repair of deterioration in concrete bridges—an overview. *Sino-British Highways and Urban Traffic Conference.* 17–22 Nov. Institution of Highways and Transportation, supplement, 6.

POCOCK D. C. and HARTLEY J. P. (1990). New techniques for diagnosis. Institution of Highways and Transportation National Workshop. *New concepts for the management of highway structures.* Leamington Spa, 20 Mar.

PRICE W. I. J. (1986). Research techniques. *Bridge inspection techniques and maintenance and repair methods.* PTRC autumn course.

PULLAR-STRECKER P. (1987). *Corrosion damaged concrete. Assessment and repair.* CIRIA and Butterworths.

PULLEN D. A.W. and CLAYTON R. F. (1981). The radiography of Swaythling Bridge. *British J. Non-destructive Testing,* **23**, No. 5, 227–232.

RHODES J. R., STOUT J. A., SIEBERG R. D. and SCHINDLER J. S. (1980). *In situ determination of the chloride content of Portland cement concrete bridge decks.* PB 275 376 Federal Highway Administration, Washington DC, USA.

SCHIESSL P. and RAUPACH M. (1993). Permanent monitoring of the corrosion risk for the reinforcement of concrete bridges. *Proc. 2nd Int. Conf. on Bridge Management.* University of Surrey, Guildford, 18–21 Apr.

SKOVESENDE S. T. (1989). Evaluation of condition surveys on concrete structures: Discussion of measurements of half-cell potentials, resistivity and related parameters. *Proc. Int. Conf. on Structural Faults and Repair.* Kensington, London, 27–29 June.

SMITH L. M. (1988). *Corrosion of reinforcement.* Central Laboratories Report 88–A9405, New Zealand.

STANDING COMMITTEE ON STRUCTURAL SAFETY (1977). 1st Report published by Construction Industry Research and Information Association (CIRIA), London.

STANDING COMMITTEE ON STRUCTURAL SAFETY (1979, 80, 82, 85, 87, 89 and 92). 3rd–8th Reports published by Construction Industry Research and Information Association (CIRIA), London.

STRATFULL R. F. (1969). *Chlorides and bridge deck deterioration.* Interim Report No. M&R 6351164, California Department of Transportation, Nov., 15.

STRATFULL R. F. (1973). *Corrosion autopsy of a structurally unsound bridge deck.* Highway (Transportation) Research Board No. 433, 1–11.

SURYAVANSHI A. K. and NAYAK B. U. (1990). Half-cell potential corrosion-monitoring technique: a review. *The Indian Concrete Journal*, Apr., 198–202.

TILLY G. P. (1990). A. S. Nowak (ed.). Evaluation of bridges-European perspective. *Bridge evaluation, repair and rehabilitation.* Nato ASI Series E: Applied Sciences. Vol. 187, ISBN 0 7923 0999 5, 367–377.

VAN DAVEER J. R. (1975). Techniques for evaluating reinforced concrete bridge decks. *J. Am. Concr. Inst.*, Dec., Title No. 72–47c.

VASSIE P. R. (1980). *A survey of site tests for the assessment of corrosion in reinforced concrete.* TRRL Laboratory Report 953.

VASSIE P. R. (1991). *The half-cell potential method of locating corroding reinforcement in concrete structures.* TRRL Application Guide 9.

WATSON A. (1963). Measurement and control of moisture content by microwave absorption. *Int. Symp. on Humidity and Moisture.* Washington DC, USA, May.

WENNER F. (1915). *A method of measuring earth resistivity.* Bulletin of the Bureau of Standards, 12, 469–478.

WIGGENHAUSER H., WILSCH G. and SCHAURICH D. (1993). Element analysis of building surfaces by laser spectroscopy. *Proc. Int. Conf. on NDT in Civil Engineering.* University of Liverpool, 14–16 Apr.

WINNET M. A. (1985). *Microprocessor controlled multiple half-cell measurements.* TRRL Application Guide AG3.

WOLFF R. (1990). Prestressing with fibre composite materials and monitoring of bridges with sensors. *Proc. 1st Int. Conf. on Bridge Management.* University of Surrey, Guildford, 28–30 Mar., 395–402.

WOLFF R. and MIESSELER H. J. (1990). A. S. Nowak (ed.). Applications with optical fibre sensor system for monitoring prestressed concrete structures. *Bridge evaluation, repair and rehabilitation.* Nato ASI Series E: Applied Sciences, Vol. 187, ISBN 0 7923 0999 5, 119–132.

WOLFF R. and MIESSELER H. J. (1993). Experience with permanent monitoring of the bridges Schiessbergstrasse and Nötsch. *Proc. 2nd Int. Conf. on Bridge Management.* University of Surrey, Guildford, 18–21 Apr.

WOOD J. G. M. (1990). G. Somerville (ed.). Predicting future decay in concrete structures. British Group of IABSE Colloquium, Cambridge, July. *The design life of structures.* Blackie Academic and Professional (an imprint of Chapman and Hall), Glasgow, 1992.

WOODWARD R. J. (1979). Application of non-destructive testing techniques to concrete bridges. *Proc. Seminar at PTRC Summer Annual Meeting, on Highway Construction and Maintenance*, University of Warwick, 9–12 July, 33–43.

WOODWARD R. J. (1981). *Conditions within ducts in post-tensioned prestressed concrete bridges.* TRRL Laboratory Report 980.

WOODWARD R. J. and WILSON D. L. S. (1991). Deformation of segmental post-tensioned precast bridges as a result of corrosion of the tendons. *Proc. Instn Civ. Engrs*, Part 1, **90**, Apr., 397–419.

ZOUBOFF V. (1973). The use of a Betatron in the radiography of road structures. Bulletin de Liaison LCPC, No. 64, Mar./Apr.

4

Repair techniques

4.1 Evolution of repair techniques

4.1.1 Bridge decks in the USA

Stewart (1972) discussed problems in the repair of bridge decks damaged by deicing salt, treating restoration as a cost-benefit expediency for gaining additional deck service life.

Clear (1974) reported on the evaluation, for permanent repair, of chloride-contaminated concrete bridge decks. To decide how much concrete to replace, he suggested a threshold chloride concentration of 1.0 lbCl^- per cubic yard (typically rather less than 0.2% by weight of cement) as a criterion for total chloride concentrations to be weighed with other information (3.3.4). Bridge decks were categorized according to whether the number of half-cell potentials more negative than −350 mV was:

(*a*) 40% or more of all readings
(*b*) between 5 and 40% or
(*c*) less than 5% and no spalling.

These criteria were tentative and subject to variation by the bridge authorities in the light of experience.

Manning and Ryell (1980) presented decision criteria for the rehabilitation of concrete bridge decks.

4.1.2 Patch repairs

The Cement and Concrete Association (1974) promulgated advice on the repair of spalled concrete using ordinary or rapid hardening Portland cement. Repair techniques for bridges generally followed traditional building practice. Portland cement mortar or fine concrete was proposed, either alone or with other materials, to make good spalled areas with the clear advantage that it is cheap and essentially similar to the concrete being repaired. The Cement and Concrete Association booklet also covered epoxide resins which offer good strength, adhesion and impermeability but have thermal expansion different from concrete and some of the hardeners are toxic. Polyester resins were given as an alternative, intermediate in cost but with other disadvantages. Natural or synthetic (styrene-butadiene) rubber latices or PVA, polyvinyl acetate, could be used either as bonding agents or to modify the properties of cement mortar. The booklet stressed the importance of removing all loose material and thoroughly cleaning the concrete and steel surfaces before placing the repair mortar. Epoxy resin is useful for a patch repair where the cover depth is low and it is not acceptable to add any extra thickness.

4.1.3 Treatment of cracks

The Cement and Concrete Association (1974) described procedures for sealing cracks, depending on the cause of the cracking and whether or not further movement was expected. If cracks were caused by corrosion of reinforcement, all defective concrete had to be cut out, the steel thoroughly cleaned and the patch repair completed. Where no further movement was expected, neat cement grout, flexible latex or more rigid epoxy resin systems were proposed to fill cracks and small cavities with the aim of matching the strength or flexibility required. Where further movement was expected, the recommendation was to cut a chase wide enough for strains to be sufficiently small for an elastic sealer to be effective.

Tabor (1979) provided much practical advice on the application of appropriate epoxy resin formulations for crack injection, mortar patching or bonding new to old concrete. Cracks may be injected by pressure, gravity or vacuum, the last method being especially suitable for honeycombed or porous concrete as it needs no shuttering, only polythene sheeting. The epoxy resins generally achieved better penetration than the cementitious grouts.

4.1.4 Sprayed concrete

Sprayed concrete (gunite or shotcrete) is generally a suitable repair technique for large areas. Because the skill of the nozzleman is all-important, the Cement and Concrete Association (1974) strongly recommended the use of experienced specialist contractors. Malmberg (1990) presented the advantages of shotcrete, plain or reinforced with steel fibres or polypropylene fibres, for repairs, especially if tensile strength or abrasion resistance was needed. (See also 6.5.3 and 7.4.4.)

4.1.5 Addressing the special problems of bridges

The OECD (1981) reported current practice in surface repairs and injection of cracks and cavities and (1983) recorded the view that the problem of repairing chloride-contaminated concrete was unresolved. The OECD (1983) gave guidance on bonding external reinforcement and on supplementary prestressing.

The Building Research Establishment (1982) proposed the same methods with a good chance of success where corrosion was due to carbonation in the absence of chlorides. Where chlorides were present, such repairs would buy some time but not long-term durability.

Manson *et al.* (1983) investigated the cost effectiveness of various overlays and impregnation techniques for the long-term rehabilitation of salt-contaminated bridge decks which had not been waterproofed. Mirza (1989) and Sawada (1990) described repair mortars for the same purpose (see 4.3.3).

Hirosawa *et al.* (1984) described seismic repair techniques for reinforced concrete columns and shear walls, developed from a United States–Japan co-operative programme.

The Concrete Society (1984) in association with FERFA, the Federation of Epoxy Resin Formulators and Applicators, issued a comprehensive report on diagnosis, the choice of method and current repair systems. Regarding patch repairs to chloride-contaminated concrete, the extent and rate of further damage nearby could not be predicted, but a resin system offered the advantage of forming a barrier against electrochemical interaction.

The Cement and Concrete Association (1985) updated their booklet with more references, including guidance from FERFA, the Concrete Society and the Building Research Establishment.

The Swedish National Road Administration (1985) issued interim requirements for the repair of concrete bridges. This covered assessment of damage, removing concrete, preparation, casting replacement concrete and the injection of cracks with epoxy. Concrete containing 0.5% or more of chloride by weight of cement had to be removed and, on removal, the aim was to remove concrete with chloride concentrations greater than 0.2% by weight of cement.

Bickley *et al.* (1986) reported to the Roads and Transportation Association of Canada on repair techniques for concrete bridge components other than bridge decks for which Manning and Ryell (1980) had already developed a general strategy. Details were given of cementitious and resin-based repair mortars; also crack and delamination injection, concrete replacement and shotcrete. How much chloride-contaminated concrete to remove remained a decision to be made in each particular case in the light of comparative costing and engineering judgement.

The Department of Transport (1986a) issued advice which was updated (1990) requiring:

(*a*) the cause of cracking to be dealt with,
(*b*) removal of cracked concrete where chloride concentration at the reinforcement exceeds 0.3% by weight of cement,
(*c*) static cracks to be sealed and impregnated if necessary,
(*d*) removal of concrete where chloride concentration at the reinforcement exceeds 0.3% by weight of cement and half-cell potentials are more negative than −350 mV,
(*e*) thorough cleaning of reinforcement by high pressure water jetting or wet blast cleaning,
(*f*) replacement of severely pitted reinforcement,
(*g*) replacement concrete to comply with the Department's Model Specification (1986b).

Keer and Emberson (1986) reported that resin-injected concrete was structurally satisfactory but research by Calder (1986) and Calder and Thompson (1988) has shown that resin injection does not effectively seal cracks against chloride ingress.

Vassie (1987) published research showing the ineffectiveness of cleaning procedures on pitted reinforcement.

PIARC Technical Committee on Bridges (1987) reviewed the experience of 15 member countries in rehabilitation of bridge decks, tabulating materials and procedures.

Mays and Wilkinson (1987) reported on structural performance of patch repairs using:

- polyester resin mortar
- epoxy resin mortar with bonding agent
- epoxy resin mortar without bonding agent
- ordinary Portland cement mortar
- cementitious mortar modified with styrene butadiene rubber polymer latex and SBR bonding agent.

With one exception in which the patch repair extended into the shear span, the ultimate strength of the repaired test beams exceeded that of unrepaired controls. Deflections, however, were increased, reflecting the lower elastic modulus of the repair materials compared with the original concrete. The bond at the repair/substrate interface is of paramount importance and the use of a primer with epoxy repairs improved their efficiency.

4.1.6 Repair options

Boam (1988) reviewed current experience of investigation and repair for which the following options were proposed:

- do nothing
- reduce corrosion rate
- repair visible defects
- carry out major repairs
- apply cathodic protection
- replace affected elements.

Bijen (1989) discussed the advantages and disadvantages of:

- traditional cement/sand mortars
- concrete casting
- shotcrete
- polymer modified cement mortars
- polymer mortars.

Bennison (1989) discussed the durability aspects of Portland cement, porosity and permeability. He described in detail the modification of Portland cement pastes by various polymers, condensed silica fume and pulverized fuel ash and the application of these materials to concrete repairs.

Wood *et al.* (1989) reviewed the position and concluded:

- repair like with like
- do not accept inadequately defined materials
- do not accept materials with irrelevant ingredients
- tailor repair mix by enhancement of original mix
- do not damage substrate when preparing it for repair
- ensure electrochemical stability of repair.

The Concrete Repair Association was formed in 1989 by a number of UK concrete repair contractors who wished to demonstrate high standards, to promote and to develop concrete repair practice.

Fisher and Menzemer (1990) reported Canadian repair practice, dealing with many bridges more than 50 years old of different types and suffering various degrees of corrosion.

Ingvarsson (1990) reported very favourably on the removal of damaged concrete by water jetting, a technique considered further in 4.3.1. Experience in Sweden found the technique:

- to be sound environmentally
- to be time saving
- to be reasonable in cost
- producing no cracking of residual concrete
- leaving reinforcement undamaged and cleaned
- providing for selective removal of concrete.

Emberson and Mays (1990a) reported on physical and mechanical properties of the following patch repair systems:

- Resinous mortars
  A Epoxy
  B Polyester
  C Acrylic
- Polymer-modified cementitious materials
  D Styrene butadiene rubber
  E Vinyl acetate
  F Magnesium phosphate
- Cementitious materials
  G OPC/sand mortar
  H HAC mortar
  I Flowing concrete

Samples were selected to establish typical characteristics but it should be noted that, in some generic categories, there are hundreds of formulations so that the results presented are not limiting values.

Emberson and Mays (1990b) reported interim conclusions on the significance of property mismatch in repair systems for structural concrete and (1990c) in axially loaded members.

Thompson (1990) observed that traditional repair, removing loose and spalling concrete and making good with Portland cement mortar or concrete, had its origins when most corrosion was due to carbonation. It is generally satisfactory in the absence of chlorides, but both Brown (1986) and Wallbank (1989) found that carbonation is not a serious problem in modern bridges. If it caused spalling, this was generally superficial and the repairs were usually without structural significancc. On the other hand, chloride contamination is likely to require the removal of comparatively large areas of concrete, debonding of reinforcement and therefore the need for temporary support. Techniques for evaluating half-cell potential and other test data were being developed by Gower and Calder (1989).

Thompson (1990) identified three possible objectives:

- restore as built, the least likely;
- conserve as found;
- permit slow continuing deterioration.

The Fédération Internationale de la Précontrainte (FIP) (1991) issued guidance on repair methods and materials, including the information due to Schiessl (1985) reproduced as Figure 4.1. The Guide also deals with supplementary prestressing and temporary support.

Cope's textbook (1987) has a chapter by Bennison on repair and protection of concrete bridges. Among various techniques, is the suggestion that fine cracks,

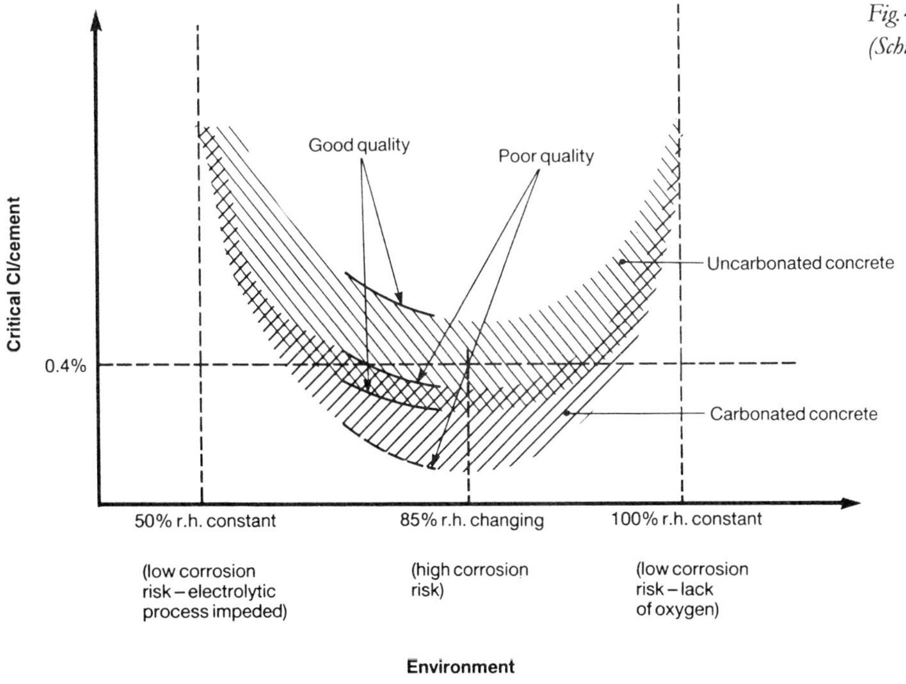

Fig. 4.1. Critical chloride content (Schiessl, 1985, and FIP, 1991)

down to 15 μm wide, whether structural or non-structural, should be repaired with low viscosity epoxy grouts.

Trout (1989) considered that the tools, technology and expertise for epoxy resin injection had become widely available and the field is no longer dominated by the few with esoteric skills. In the USA, he saw over 70% of crack injection as preventive maintenance to protect steel reinforcement from corrosion and he claimed that virtually any visible crack is injectable provided it is not impacted with contaminants.

The American Concrete Institute Committee 224 (1984) summarized the causes of cracking and the principal techniques for repair. The treatment of cracking is considered further in the light of wider thinking.

4.2 Treatment of cracks

4.2.1 Corrosion-induced cracking

Cracking, usually along the line of the reinforcement, caused by corrosion cannot be arrested by injection. The corroding steel must be exposed, concrete removed to at least the aggregate size behind the bar, surfaces of steel and concrete cleaned and the cavity reinstated.

4.2.2 Cracking due to alkali-silica reaction

The Department of Transport (1990a) set out advice for the management of structures in which ASR has been diagnosed. It may be appropriate to monitor crack widths and/or impregnate the concrete.

4.2.3 Progressive cracking

If monitoring shows that a crack is widening or lengthening, the structural cause of this has to be found and remedied. The FIP (1986) gave a classification of cracks in relation to their primary causes.

4.2.4 Injection of stabilized cracks

The objective is:

- to restore structural integrity across the crack and/or
- to block access of water and chloride ions to the steel.

The crack should be as clean as possible before filling. Crack filling is mostly done by pressure injection but in some cases by ponding and gravity. Swedish National Road Administration (1985) required low viscosity epoxy injection fluid to be used at pressures between 0.1 and 0.3 N/mm^2 and, for good results, the technique was only to be used for crack widths of 0.2 mm or more. Higher pressures were permitted, provided that the crack width was continuously monitored and pressure reduced if the width increased by more than 0.2 mm. The FIP (1991) drew the distinction between low pressure (up to about 2 N/mm^2) and high pressure injection (up to 30 N/mm^2). The pressure is selected to give a rate of penetration to suit the particular resin. The vacuum technique may offer some advantages, especially for highly cracked or porous surfaces.

The OECD (1989) reports the general use of low viscosity polymers and resins, acrylic resins for submerged structures, and also suggests that cementitious grouts may be used for cracks wider than 1 mm.

Calder (1986) reported on the structural performance of cracked slabs (see 6.2.2).

Calder and Thompson (1988) investigated the durability of slabs repaired with epoxy, polyester and methylmethacrylate resins (see 6.2.3).

Calder (1990) reported on the protection afforded by various crack injection techniques (see 6.2.4).

4.2.5 Non-progressive active cracks

Structural cracks may approximate to the design width under maximum loading and return to some lesser width under lesser loading. Normally these are left untreated as they do not usually pose a durability problem. If, however, there is concern, for instance, about chloride contamination reaching the steel along such a crack, consideration may be given to sealing it. The prospects of success are not high with a sealing material subjected to excessively high strains.

The Concrete Society (1984) suggested that an ultimate remedy could be to cut the surface adjacent to the crack, forming a slot to receive a low modulus joint sealant. If possible the filler or sealant should be applied when the crack is at its widest.

The OECD (1989) suggested polyurethane resins, provided the application is not wet, and the use of an acrylic gel for submerged work.

4.2.6 Crack restraint

Stratton, Alexander and Nolting (1977 and 1982) reported a technique consisting of sealing the crack, drilling holes across the plane of cracking and bonding new reinforcement with low pressure epoxy grout to prevent further movement.

Ramey (1983) reported the repair of continuous reinforced concrete bridge girders which were broken at the reinforcement splice locations near the dead load inflection points during a lifting operation. The reinforcement splices were welded and integrity of the girders fully restored by epoxy injection. This was proved by evaluating the responses of the bridge jacking before the damage, one month after repair and one year after repair.

However, Collins and Roper (1990), investigating alternative shear crack repair techniques, found that, although stitching achieved some local stiffening, the stitch-anchorages created new stress concentrations and the effective strengthening was very little compared with the other methods tested. The best method was vertical post-tensioning. The other two techniques which were not quite as effective were:

- resin injection of the crack and
- bonding external stirrups.

4.3 Reinforced concrete repair

4.3.1 Concrete removal

If structural integrity will be affected, temporary support may have to be provided. Edges of the repair area should be saw cut to a depth of 10 to 25 mm and all loose, defective and contaminated concrete removed to prescribed limits, leaving the substrate sound and without fractures. Swedish National Road Administration (1985) prescribed detailed requirements for either compressed air chipping or hydrodemolition, edges to have a 60–90° angle to the reinforcement or the surface.

Bickley *et al.* (1986), in their Report to the Roads and Transportation Association of Canada, suggest the use of light jackhammers or water blasting.

New Civil Engineer (1986) described a water jetting lance capable of cutting out concrete at about 1 m^3 per day and perhaps £350 per m^3—40% quicker but 60% more expensive than pneumatic breaking.

PIARC (1987) Technical Committee on Bridges, reviewing methods used in 14 countries, found that all used pneumatic hammers, some already used hydrodemolition widely and others were introducing this technique. Ranges of productivity and costs are given in Table 4.1.

Table 4.1. Productivity and costs

Method	Productivity: m^3/h	Costs £/m^3
Pneumatic hammers	0.015–0.5	1200–200
Hydrodemolition	0.5–1.5	600–200

There was consensus that hydrodemolition could remove defective material more selectively than other methods such as bushhammers, milling, other mechanical methods or the very seldom used thermal methods.

Silfwerbrand (1988) reported that the water jet technique can result in such a strong bond between old and new concrete that the composite repaired slabs can have ultimate strength comparable with slabs of similar depth cast in one pour.

Anthony (1988) described Conjet, a robot controlled water jet operating at 100 to 120 N/mm^2 capable of removing 10 m^2 of concrete per hour. Nilsson (1989) claimed that Conjet is 50% cheaper and faster than conventional concrete breaking with superior results.

Ingvarsson and Eriksson (1988) comprehensively described hydrodemolition technique using pressures up to 400 N/mm^2.

Highways (1989) reported the use of the technique on Midland Links viaduct crossbeams.

Silfwerbrand (1990) investigated the structural interaction of concrete repair materials with concrete substrates prepared by jackhammers, sandblasting and water jetting. Static loading tests on composite slabs showed similar failure loads for the three methods. However, pull-off tests showed that the bond at the interface was better for the water-jetted interface, with only 7% failures compared with 31% for the jackhammered and 38% for the sandblasted interfaces. The mean failure stress for the water-jetted specimens was about 2.2 N/mm^2 compared with about 1.7 and less than 1.0 for the sandblasted and jackhammered specimens respectively. The most important difference between the methods was that the jackhammers caused microcracking and the others did not. In other tests, water jetting followed by shotcrete repairs were effective provided the interface did not coincide with the reinforcement plane. This technique was especially suited to vertical or overhead surfaces, avoiding the need for formwork for poured concrete.

The Department of Transport (1990) pointed out the need to ensure structural integrity by staging the work or by propping. High pressure water jetting is suggested for the removal of large areas of concrete where appropriate and with necessary safety arrangements.

Medeot (1990) described the technology of hydrodemolition, developed at the end of the 1970s in Italy, and gave many examples of practical applications.

New Civil Engineer (1991) examined some commercial problems with hydrodemolition in the UK, putting a typical cutting rate for hand held equipment at about 0.3 m^3 per day but very much higher rates were achievable with automation.

Manning (1991) gave a synthesis of procedures and equipment used in the States to remove deteriorated concrete from highway bridges.

New Civil Engineer (1991) described pier replacement, with water jetting at 248 N/mm^2 costing £2000 per m^3, reckoned to be twice as quick and half the price of patching.

New Civil Engineer (1991) described automated equipment taking out 800 m^3 of deck concrete at up to 1 m^3 per h.

The FIP (1991) gave information on:

- hydraulic water jetting (10–40 N/mm^2)
- high pressure water jetting (40–120 N/mm^2)
- hydro-jetting (140–240 N/mm^2)

with comprehensive advice on other mechanical methods and other aspects of preparing concrete surfaces. The damage and dangers associated with thermal methods and chemical methods were also outlined.

Schrader (1992) advised against mechanically roughening, wetting or applying grout or mortar to the concrete surface to improve bond. He found that latex emulsions could reduce rather than improve bond.

4.3.2 Reinforcement

Vassie (1987) studied the effect on durability of steel cleaning procedures. Neither manual nor powered wire brushing of locally corroded steel will prevent corrosion from going on in the repair. Grit blasting is effective on coarsely pitted and on generally corroded steel surfaces. This supports the Concrete Society recommendation (1984) to clean chloride-contaminated steel by grit blasting or by very high pressure water jetting. (See also 5.3.4.)

Pritchard and Chubb (1987) suggested that the loss of strength of a bar due to corrosion may actually be less than at first appearance for two reasons:

(*a*) the rather slow loss of section due to general corrosion,
(*b*) strain hardening if a pitted bar is stressed to yield.

Tests on pitted bars had showed that up to 30% loss of section did not lower the yield strength based on the gross section and that 60% loss of section only reduced it by 10%. They had found that loss of concrete section by spalling and reduction of bond strength were more significant effects.

The Department of Transport (1990) requires reinforcement to be cleaned by high pressure water jetting or wet blasting or replaced if it is extensively pitted or has suffered significant loss of section.

4.3.3 Repair material

Perkins (1986) dealt mainly with non-structural repair and, for concrete bridges, recommended the use of polymers, mainly SBR but also acrylics.

Bennison (1989) discussed the durability of concrete and the design of the following types of repair material:

- unmodified cementitious
- polyester and epoxy resins
- polymer-modified cementitious
- cement/pozzolanic-modified materials.

The choice is made to suit the required combination of

- mechanical properties akin to parent concrete,
- adhesion in dry, damp or wet conditions,
- low shrinkage in curing and long term.

Maddocks (1989) described the repair of concrete bridges with special reference to hand-placed mortar, including the choice of mortar.

Mirza (1989) reported laboratory tests on the performance of 34 types of overlay mortar compared with that of a standard ASTM cement mortar. Cementitious grouts and polymer-modified cement based mortars bonded better than the standard reference mortar but not so well as the epoxy mortars. However, the epoxy materials were shown to be less compatible with the parent concrete in other tests.

Plum (1989) compared the performance of polymer-modified cementitious materials, unmodified concrete and polymer concretes, noting the significant effects of the curing conditions.

Sawada (1990) described 'asana refresh', a repair procedure using SBR polymer cement mortar and an organic paint film.

Since bridge concrete tends to be wet, the Department of Transport (1990) requires prebatched acrylic polymer-modified cementitious mortar to be used for minor repairs. FERFA (1991) has issued guidance on the safe handling of such formulations.

4.3.4 Replacement concrete

The Department of Transport (1986b) requires the use of blended cement concrete for bridge decks, high flow concrete for sides and soffits of beams and strict controls to minimize shrinkage cracking and produce a dense durable repair. Properties of high flow concrete are considered in 6.5.5.

4.3.5 Sprayed replacement concrete

The Cement and Concrete Association (1985) defined 'gunite' as sprayed concrete with maximum aggregate size less than 10 mm and 'shotcrete' with aggregate size 10 mm or greater. Equipment, materials, mixing, application and testing were described.

Andalen (1989) described some of the properties of shotcrete.

- For both wet and dry mixes, the best bond is obtained on dry substrates and the worst on wet substrates.
- The strength and air content of the substrate has little effect on bond.
- Of the samples tested, wet mix provided better resistance to frost and salt than dry mix.
- Dry mix results were variable due to the difficulty in controlling the water/cement ratio.

The Department of Transport (1990) suggested the use of sprayed concrete where reinforcement is not too congested. Reference should be made to the comprehensive Code of Practice published by the Sprayed Concrete Association. (See also 6.5.3.)

4.3.6 Repair of plastic hinge zones

Lawrance and Jacks (1990) reported tests on model bridge piers subjected to various levels of damage by cyclic loading and repaired using techniques appropriate to the damage. All of the repaired piers achieved peak loads and ductility in accordance with the New Zealand code, the epoxy injected (least damaged) pier having increased capacity and ductility.

4.4 External reinforcement

Where justified, supplementary reinforcement may be added to a concrete member in the form of mild steel plates bonded by a suitable adhesive to the concrete surface. The technique was applied in the UK in 1975 on two pairs of M5 motorway bridges at Quinton after investigation of cracking revealed a design error in the stopping off of reinforcing bars. Raithby (1980) reported loading tests carried out before and after plating showing an 11% increase in the deck stiffness.

Research was undertaken at TRRL and other centres to provide information on design, application and long-term durability of the structural bonding.

Mays and Tilly (1982) investigated the fatigue performance of bonded reinforcement and found it to be satisfactory.

SCOSS (1985) considered that plates should have mechanical fixings to prevent them from falling in the event of debonding and repeated their concern that performance of resins under long-term loading was not well known. However, they noted that 'most bridges receive only relatively little permanent service load and therefore creep strains may not be important'.

Calder (1988) reported the performance after ten years' exposure of small test beams made using the same resin as that used at Quinton (Resin 1). Slight corrosion had only marginal effect on failure loads.

Eberline *et al.* (1988) provided an extensive review of relevant literature.

Mays and Hutchinson (1988) defined the main requirements for a successful joint, drew the distinction between structural and semi-structural joints and proposed a compliance spectrum for a two-part cold cure adhesive.

Calder (1989) reported tests on larger beams made using Resin 1 and Resin 2. The latter was more flexible and proved to be unsatisfactory after eight years' exposure due to the penetration of moisture and the onset of corrosion which caused debonding. Bonding on the Quinton bridges has been monitored by manual impact-echo testing and has remained sound except in one location where water carrying deicing salt penetrated the interface causing corrosion.

Since the Quinton bridges, a further eight bridges in the UK have been strengthened in this way to make up various deficiencies.

SCOSS (1989) welcomed the research that had been undertaken but repeated its earlier advice that

> 'resin-bonded steel plates should only be used after careful structural appraisal of the problem and where frequent monitoring of their performance in use can be ensured and where premature failure of the resin bond can be tolerated and replacement can be effected. Resin-bonded plates should specifically not be used on concrete of unsatisfactory quality or which is deteriorating as a result, for example, of chloride or sulphate penetration or by alkali-silica reaction.'

Jones, Swamy and Hobbs (1990) summarized the technique of bonding steel plates, of width : thickness ratio at least 50 : 1, its application, design and performance under load. Research, including alternatives to mild steel, was noted.

Sonada, Matsui and Hayashi (1990) described bolting thin steel plates at measured clearance from the bridge soffit, sealing edges and then pumping in epoxy resin to make the bond.

Meier and Kaiser (1991) described research at EMPA, Dübendorf, Switzerland, into post-strengthening structures with carbon fibre reinforced plastics (CFRP).

Taylor and Dykes (1992) described plate bonding on seven inverted prestressed concrete T-beams of a 1968 pseudo-slab bridge to sustain an increase in dead load resulting from a carriageway improvement.

Papers on strengthening bridges were presented by Shaw (1993), Henwood and O'Connell (1993), Mays (1993) and Oehlers (1993).

4.5 Supplementary prestress

The OECD (1983) gave several case histories and defined two ways of applying supplementary prestressing:

- vertical or inclined tendons to augment shear resistance,
- longitudinal tendons to augment bending resistance.

Similar alternative anchorage arrangements suggested by the FIP (1991) are shown in Figures 4.2 to 4.5 and Figure 4.6 shows a deflector.

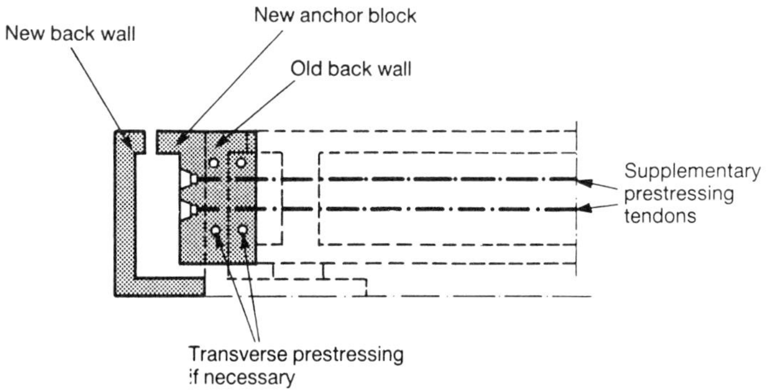

Fig. 4.2. Anchorage of supplementary prestressing at end of girder (FIP, 1991)

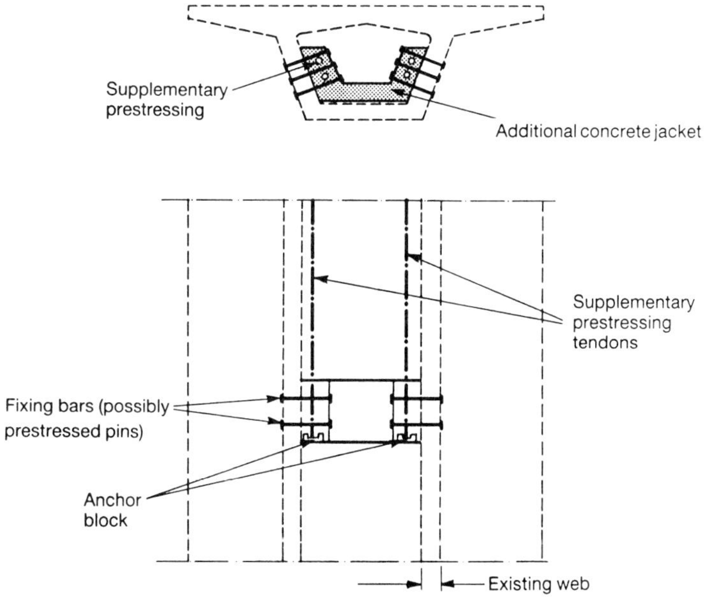

Fig. 4.3. Anchorage of supplementary prestressing with additional support (FIP, 1991)

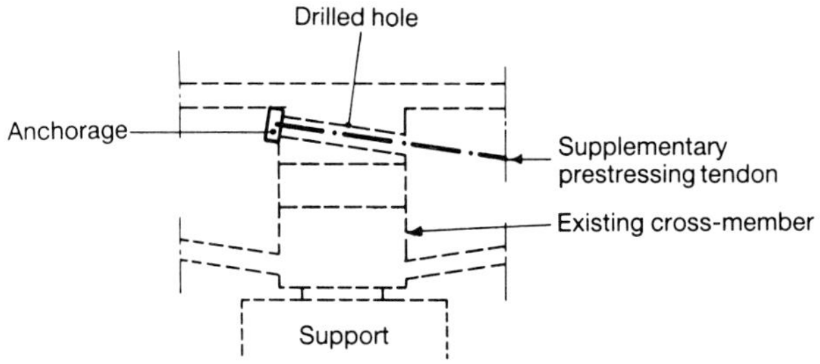

Fig. 4.4. Anchorage of supplementary prestressing at existing diaphragm (FIP, 1991)

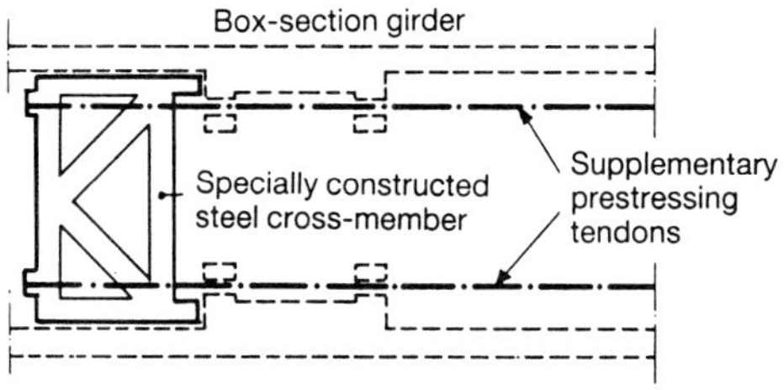

Fig. 4.5. Anchorage of supplementary prestressing with auxiliary steel frame (FIP, 1991)

Fig. 4.6. Deflector for supplementary prestressing (FIP, 1991)

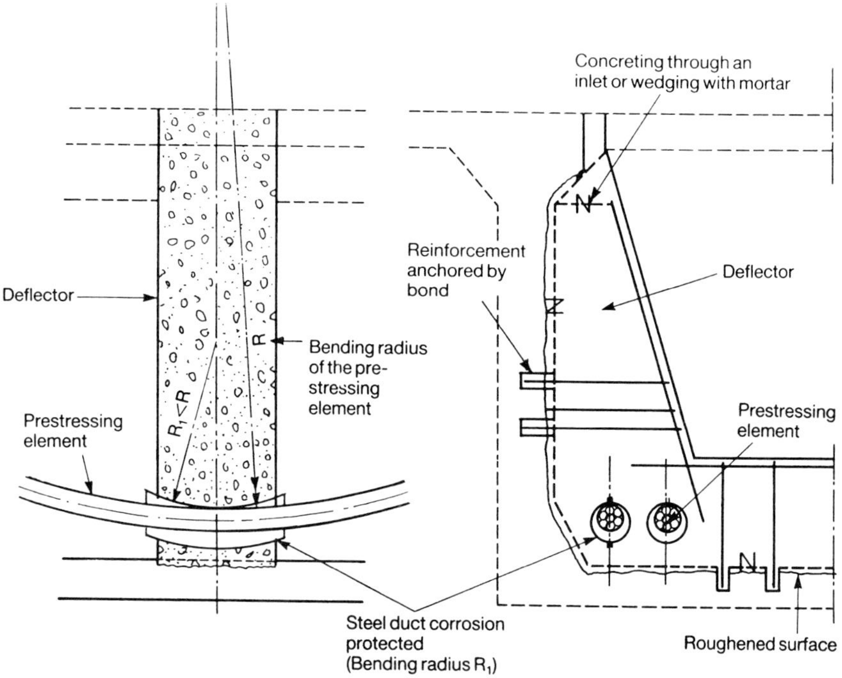

Dilger and Ghali (1985) suggested remedial measures for the cracked webs of prestressed concrete bridges.

Preston, Osborn and Roach (1987) developed two methods for repairing deteriorated 1950s prestressed concrete box beams. They required the removal of deteriorated concrete, placing either

(*a*) external epoxy-coated reinforcement or
(*b*) external epoxy-coated post-tensioned strand

and the application of gunite to the soffits. Both methods restore the ultimate flexural capacity. Method *a* does not restore prestress but is the less costly. Method *b* restores most of the lost prestress but may cause high tensile stress in the tops of beam ends. Also, epoxy coatings have come under suspicion because of failures in the USA (see 5.3.1).

Lacroix (1989) attempted to provide guidance on the pathology and repair of prestressed concrete structures such as bridges.

Trinh (1990) reported examples of laboratory and field work on the development of structural strengthening by adding external prestressing.

The FIP (1991) gave similar information with more details. The technique can be applied to reinforced or prestressed structures. The short prestressing elements may be sensitive to alignment deviations and require the use of an ancillary nose for the prestressing jack. Design is similar to that of new works except that creep losses are lower due to the age of the existing concrete. Space may not allow the provision of bursting reinforcement but the needs may be met by transverse prestressing.

The FIP also included strengthening by imposed deformation. This is achieved by the relative displacement of supports to induce a self-equilibrated stress condition. However, relief in some sections will worsen the situation in others.

4.6 Underwater repair

The Concrete Society (1990) published general information on underwater concreting including a section dealing with repair techniques and materials.

4.7 Structural aspects

When planning repairs, the following structural aspects have to be taken into account.

Loss of section of concrete or steel, whether by deterioration or by accidental damage, affects stresses.

If concrete removal will debond reinforcement, partially or completely, or render its anchorage ineffective, temporary support may be required or the work may have to be carried out in stages.

If the whole of the dead load is carried by jacks on temporary supports while the repair is carried out and until it is fully cured, the repair material should participate fully, as in a new structure. Otherwise the repair material will only be stressed by loading applied after completion of the repair.

If no propping is provided, freshly placed material may be adversely affected by transient loading. This may suggest some temporary traffic control. A study by Harsh and Darwin (1984) suggests that traffic vibrations were detrimental to bond and the quality of high flow (slump) repair concrete.

The structural effectiveness of the repair material also depends on its strength, elasticity and bond to the existing concrete.

Differential shrinkage and creep will affect the structural contribution of the repair.

Particular care should be taken when breaking out concrete in end anchorage zones and anywhere in prestressed concrete.

Notes and references on the demolition of reinforced and prestressed concrete structures are given in Appendix 3.

Prestress levels in a prestressed concrete member will not be affected by adding material, as when repairing impact damage.

Repair material can be prestressed by supplementary post-tensioning or, if it is feasible, by restressing tendons previously destressed.

4.8 References

American Concrete Institute Committee 224 (1984). Causes, evaluation and repair of cracks in concrete structures. Report No. 2241.R-84. *J. Am. Concr. Inst.*, May-June, 211–229.

Andalen A. (1989). Shotcrete for concrete repairs—function and durability. *Symp. on the Durability of Structures*. Lisbon, International Association for Bridge and Structural Engineering.

Anthony M. (1988). Water over a troubled bridge. *Contract Journal*. 13 Oct., 33.

Bennison P. (1989). Materials for concrete repair: innovation. *Proc. Int. Conf. on Structural Faults and Repair*. Kensington, London, 27–29 June.

Bickley J. A., Emery J. J., Sturrup V. and MacDonald P. (1986). *Repair techniques for concrete bridge components*. Roads and Transportation Association of Canada. ISBN 0 919 098 48 7.

Bijen J. M. J. M. (1989). Maintenance and repair of concrete structures. *Heron*, **34**, No. 2. 35–39.

Boam K. (1988). Concrete highway structures-current experience of investigation and repair. *Concrete Repair Conference*, London Tara Hotel, 16 Mar. Palladian Publications Limited.

Boyd W. K., Lankard D. R. and Slater J. E. (1979). *Influence of bridge deck repairs on corrosion of reinforcing steel*. Transportation Research Board, Washington DC, USA.

Brown J. H. (1986). The performance of concrete in practice. *A field study of highway bridges*. TRRL Contractor Report 43.

Building Research Establishment (1982). The durability of steel in concrete: Part 3. *The repair of reinforced concrete*. Digest 265.

Burley E., Rigden S. and Abu-Tair A. (1989). The performance of different repair techniques and materials under static loading. *Proc. Int. Conf. on Structural Faults and Repair*. Kensington, London, 27–29 June.

CALDER A. J. J. (1986). *Repair of cracked reinforced concrete: assessment of injection methods.* TRRL Research Report 81.

CALDER A. J. J. (1988). *Exposure tests on externally reinforced concrete beams—performance after 10 years.* TRRL Report 129.

CALDER A. J. J. and THOMPSON D. M. (1988). *Repair of cracked reinforced concrete: assessment of corrosion protection.* TRRL Research Report 150.

CALDER A. J. J. (1989). *Exposure tests on 3.5 m externally reinforced concrete beams—the first 8 years.* TRRL Research Report 191.

CALDER A. J. J. (1990). Corrosion protection afforded by various crack injection techniques. *Proc. 3rd Int. Conf. on Corrosion of Reinforcement in Concrete Construction.* Wishaw, Warwickshire, 21–24 May. Published for the Society of Chemical Industry by Elsevier Applied Science, 433–445.

CEMENT and CONCRETE ASSOCIATION (now BRITISH CEMENT ASSOCIATION) (1974). *The repair of concrete structures.* Allen R. T. L.

CEMENT and CONCRETE ASSOCIATION (now BRITISH CEMENT ASSOCIATION) (1985). *The repair of concrete structures.* Allen R. T. L.

CLEAR K. C. (1974). *Evaluation of Portland cement concrete for permanent bridge deck repair.* FHWA-RD-74–5, Federal Highway Administration, Washington DC, USA.

COLLINS F. and ROPER H. (1990). Laboratory investigation of shear repair of reinforced concrete beams loaded in flexure. *ACI Mater. J.*, Mar.–Apr., 149–159.

CONCRETE SOCIETY (1984). *Repair of concrete damaged by reinforcement corrosion.* Technical Report No. 26.

CONCRETE SOCIETY (1990). *Underwater concreting.* Technical Report No. 35.

COPE R. J. (ed.) (1987). *Concrete bridge engineering: performance and advances.* Elsevier Applied Science. ISBN 1 85166 110 7.

DEPARTMENT of TRANSPORT (1986a). *The investigation and repair of concrete highway structures.* Departmental Advice Note BA 23/86.

DEPARTMENT of TRANSPORT (1986b). *Materials for the repair of concrete highway structures.* Departmental Standard BD 27/86.

DEPARTMENT of TRANSPORT (1990). *Inspection and repair of concrete highway structures.* Departmental Advice Note BA 35/90.

DILGER W. H. and GHALI A. (1985). Remedial measures for cracked webs of prestressed concrete bridges. *J. Prestressed Concr. Inst.*, July, **19**, No. 4, 76–85, Illinois, USA.

EBERLINE D. K., KLAIBER F. W. and DUNKER K. (1981). *Bridge strengthening with epoxy-bonded steel plates.* Transportation Research Record No. 1180, Washington DC, USA, 7–11.

EMBERSON N. K. and MAYS G. C. (1990a). R. K. Dhir and J. W. Green (eds). Design of patch repairs: measurements of physical and mechanical properties of repair systems for satisfactory structural performance. *Proc. Int. Conf. on Protection of Concrete.* University of Dundee, 11–13 Sept.

EMBERSON N. K. and MAYS G. C. (1990b and c). Significance of property mismatch in the patch repair of structural concrete. *Mag. Concr. Res.*, **42**, No. 152, Sept. Part 1: Properties of repair systems, 147–160. Part 2: Axially loaded reinforced concrete members, 161–170.

FERFA (1991). *Safe handling for users of polyester/acrylic resin formulations.* Application Guide No. 7.

FIP (1986). *Inspection and maintenance of reinforced and prestressed concrete structures.* Fédération Internationale de la Précontrainte. Thomas Telford, London.

FIP (1991). *Repair and strengthening of concrete structures. Guide to good practice.* Fédération Internationale de la Précontrainte. Thomas Telford, London.

FISHER J. W. and MENZEMER C. R. (1990). A. S. Nowak (ed.). Bridge repair methods—

US/Canadian practice. *Bridge evaluation, repair and rehabilitation.* Nato ASI Series E: Applied Sciences, Vol. 187, ISBN 0 7923 0999 5, 495–512.

GOWER M. R. and CALDER A. J. J. (1989). Monitoring and evaluating reinforcement corrosion using CONTEST. *Proc. 3rd Int. Conf. on Deterioration and Repair of Reinforced Concrete in the Arabian Gulf.* Bahrain, Oct.

HARSH S. and DARWIN D. (1984). *Effects of traffic induced vibrations on bridge deck repairs.* SM Report 9, Kansas University.

HENWOOD A. M. and O'CONNELL K. J. (1993). The use of externally bonded plates to strengthen Bolney Flyover. *Proc. 2nd Int. Conf. on Bridge Management.* University of Surrey, Guildford, 18–21 Apr.

HEWLETT P. C. and MAY J. O. (1989). Third Party Certification of repair materials and systems using Agrément. *Proc. Int. Conf. on Structural Faults and Repair.* Kensington, London, 27–29 June.

HIGHWAYS (1989). High pressure water jetting delivers the punch without the punishment. *Highways*, Dec., **57**, No. 1956, 26.

HIROSAWA M., NAKATA S., GOTO T., YOSHIMURA M., SHIGA T., SHIBATA A. and IMAI H. (1984). *Development of seismic repair techniques for reinforced concrete members.* Department of Civil Engineering, Pittsburgh, USA, 597–612.

INGVARSSON H. and ERIKSSON B. (1988). Hydrodemolition for bridge repairs. *Nordisk Betong (Nordic Concrete Journal)*, No. 2–3.

INGVARSSON H. (1990). Hydrodemolition for bridge repair. *British–Swedish Seminar on Concrete Repairs.* 10–11 Dec., P-90:11. Chalmers University of Technology, Goteborg.

JONES R., SWAMY R. N. and HOBBS B. (1990). Bridge strengthening using epoxy bonded steel plates. *Highways*, July, **58**, No. 1963, 23–25.

KEER J. G. and EMBERSON N. (1986). *Resin injected reinforced concrete members under load: a theoretical and experimental study.* TRRL Contractor Report 31.

LACROIX R. (1989). Renovation of prestressed concrete structures. *New Zealand Concrete Construction*, **33**, 26–29.

LAWRANCE G. M. and JACKS D. H. (1990). *Repair techniques for plastic hinge zones in bridge piers.* Central Laboratories Report 90–B5410. Lower Hutt, New Zealand, 65.

MADDOCKS J. (1989). Repair and refurbishment of reinforced concrete bridges—technically correct approach. *Proc. Int. Conf. on Structural Faults and Repair.* Kensington, London, 27–29 June.

MALMBERG B. (1990). Dry mix material for concrete repair. *British–Swedish Seminar on Concrete Repairs.* Chalmers University of Technology, Goteborg.

MANNING D. G. and RYELL J. (1980). *Decision criteria for the rehabilitation of concrete bridge decks.* Research Record 762 Transportation Research Board, Washington DC, USA.

MANNING D. G. and RYELL J. (1981). A strategy for the rehabilitation of concrete bridge decks. *RTAC Forum*, **3**, No. 4.

MANNING D. G. (1991). *Removing concrete from bridges.* NCHRP Synthesis of Highway Practice 169. Transportation Research Board, Washington DC, USA.

MANSON J. A., SLUTTER R. G. and VANDERHOFF J. W. (Lehigh University) and WEYERS R. E., CADY P. D., KLINE D. E., MOORE G. R., WILSON D. T. and WONG K. H. (Pennsylvania State University) (1983). *Long-term rehabilitation of salt-contaminated bridge decks.* National Cooperative Highway Research Program Report 257. Transportation Research Board, Washington DC, USA.

MAYS G. C. and TILLY G. P. (1982). Long endurance fatigue of bonded structural joints. *J. Adhesion and Adhesives*, **2**, Apr., 109–113.

MAYS G. C. and WILKINSON W. (1987). Polymer repairs to concrete: their influence on structural performance. *Katharine and Bryant Mather Int. Conf. on Concrete Durability.*

Detroit, ACI, SP-100.
MAYS G. C. and HUTCHINSON A. R. (1988). Engineering property requirements for structural adhesives. *Proc. Instn Civ. Engrs*, Part 2, **85**, 485–501.
MAYS G. C. (ed.) and TABOR L. J. (1992). Repair materials and techniques. *Durability of concrete structures: investigation, repair and protection*. E. and F. N. Spon, 82–129.
MAYS G. C. (1993). The use of bonded external reinforcement in bridge strengthening: requirements of the adhesive. *Proc. 2nd Int. Conf. on Bridge Management*. University of Surrey, Guildford, 18–21 Apr.
MEDEOT R. (1990). Hydrodemolition—a modern technique of concrete removal in bridge repair. *Proc. 1st Int. Conf. on Bridge Management*. University of Surrey, Guildford, 28–30 Mar., 765–776.
MEIER U. and KAISER H. (1991). Strengthening of structures with CFRP laminates. *Advanced Composites Materials in Civil Engineering Structures Proc.* MT Division. American Society of Civil Engineers, Las Vegas.
MIRZA J. (1989). Compatibility of various overlays with parent concrete. *Proc. Int. Conf. on Structural Faults and Repair*. Kensington, London, 27–29 June.
NATO WORKING GROUP REPORT IV (1990). A. S. Nowak (ed.). Bridge repair. *Bridge evaluation, repair and rehabilitation*. Nato ASI Series E: Applied Sciences, Vol. 187, ISBN 0 7923 0999 5.
NEW CIVIL ENGINEER (1983). Swiss work to beat bridge decay. *New Civil Engineer*, 17 Nov. 1983.
NEW CIVIL ENGINEER (1986). Water jets into new technology. *New Civil Engineer*, 24 July 1986, 18–19.
NEW CIVIL ENGINEER (1991). Under pressure. *New Civil Engineer*, 28 Feb. 1991, 22.
NEW CIVIL ENGINEER (1991). Pier pressure. *New Civil Engineer*, 1 Aug. 1991, 16–17.
NEW CIVIL ENGINEER (1991). A cut above. *New Civil Engineer*, 15 Aug. 1991, 22.
NILSSON L. (1989). Conjet—an efficient demolition tool. *Construction Repair*, Mar., **3**, No. 2, 40–42.
OECD (1981). *Bridge maintenance*. Road Research Group Report. Organisation for Economic Co-operation and Development, 2, rue André Pascal, 75775 Paris Cedex 16.
OECD (1983). *Bridge rehabilitation and strengthening*. Road Transport Research Group Report. Organisation for Economic Co-operation and Development, 2, rue André Pascal, 75775 Paris Cedex 16.
OECD (1989). *Durability of concrete road bridges*. Road Transport Research Group Report. Organisation for Economic Co-operation and Development, 2, rue André Pascal, 75775 Paris Cedex 16.
OEHLERS D. J. (1993). Strengthening bridge corbels using bonded plates. *Proc. 2nd Int. Conf. on Bridge Management*. University of Surrey, Guildford, 18–21 Apr.
PERKINS P. H. (1986). *Repair, protection and waterproofing of concrete structures*. Elsevier Applied Science. ISBN 1 8516 008 9.
PIARC TECHNICAL COMMITTEE ON BRIDGES (1987). *Repair and Rehabilitation of Bridge Decks. (Section 4.)* Permanent International Association of Road Congresses, 27, rue Guenegaud, 75006 Paris.
PLUM D. R. (1989). Polymer modified materials and the curing environment. *Proc. Int. Conf. on Structural Faults and Repair*. Kensington, London, 27–29 June.
PRESTON H. K., OSBORN A. E. N. and ROACH C. E. (1987). *Restoration of strength in adjacent prestressed concrete box beams*. Pennsylvania Department of Transportation, 122.
PRITCHARD B. P. and CHUBB M. S. (1987). Concrete bridge integrity assessment. *Structural assessment—the use of full and large scale testing*. Butterworth, 9–19.
RAITHBY K. D. (1980). *External strengthening of concrete bridges with bonded steel plates*. TRRL

Research Report SR 612.
RAMEY G. E. (1983). *An experimental evaluation of the adequacy of a rebar splice weldment/epoxy injection repair procedure for concrete girders.* Final Report. Auburn University, Alabama; Alabama State Highway Department; Federal Highway Administration, 78.
SAADATMANESH H. and EHSANI M. R. (1990). Fiber composite plates can strengthen beams. *Concrete International: Design and Construction*, **12**, No. 3, 65–71.
SAWADA E. (1990). Repair method for salt-damaged reinforced concrete structures. *Concrete International: Design and Construction*, **12**, No. 3, 37–41.
SCHIESSL P. (1985). *Protection of concrete.* B-4.1. Bulletin 166, Comité Euro-International du Béton.
SCHRADER E. K. (1992). *Mistakes, misconceptions and controversial issues concerning concrete and concrete repair.* Part 3 of a three-part series. Concrete International.
SETHIA M. R. and MODI K. K. (1982). Sprayed concrete in repair and strengthening of old bridges. *Indian Highways*, Mar., **10**, No. 3, 35–38.
SHARMA A. K. and ANDREWS G. (1989). Testing of repaired reinforced concrete beams. *Proc. Int. Conf. on Structural Faults and Repair.* Kensington, London, 27–29 June.
SHAW M. (1993). Strengthening bridges with externally bonded reinforcement. *Proc. 2nd Int. Conf. on Bridge Management.* University of Surrey, Guildford, 18–21 Apr.
SILFWERBRAND J. (1988). Old and new concrete acting together at bridge repairs. *Nord Betong*, (Stockholm), No. 5, 3–7.
SILFWERBRAND J. (1990). Structural interaction between concrete and concrete repair materials. *British–Swedish Seminar on Concrete Repairs.* Chalmers University of Technology, Goteborg.
SONADA K., MATSUI S. and HAYASHI H. (1990). Epoxy bonded steel plate method for repairing deteriorated slabs. *International Association for Bridge and Structural Engineering Symposium*, Brussels. Mixed structures, including new materials. IABSE Reports, Vol. 60, 701 *et seq.*.
STANDING COMMITTEE ON STRUCTURAL SAFETY (1978, 1985 and 1989). 2nd, 6th and 8th Reports published by Construction Industry Research and Information Association (CIRIA), London.
STEWART C. F. (1972). *Bridge deck restoration methods and procedures.* Part 1. Repairs. PB214–635, 44.
STRATTON F. W., ALEXANDER R. and NOLTING W. (1977). *Cracked structural concrete repair through epoxy injection and rebar insertion.* FHWA-KS-RD-76–2, Kansas Department of Transportation.
STRATTON F. W., ALEXANDER R. and NOLTING W. (1982). *Cracked structural concrete repair through epoxy injection and rebar insertion.* FHWA-KS-82–1, Kansas Department of Transportation.
SWEDISH NATIONAL ROAD ADMINISTRATION (1985). *Repair of concrete bridges.* TB 151. Interim edition 20 Mar.
TABOR L. J. (1979). Concrete repairs. *Precast Concrete*, Feb., **10**, No. 2, 65–68.
TAYLOR D. C. and DYKES J. (1992). Plate bonding—a Scottish case study. *J. Instn Highw. and Transpn*, July, 26–28.
THOMPSON D. M. (1989). Effectiveness of repair of reinforcement corrosion in a motorway viaduct. *Proc. Int. Conf. on Deterioration and Repair of Reinforced Concrete in the Arabian Gulf.* Bahrain Society of Engineers, 157–171.
THOMPSON D. M. (1990). Concrete repair in highway bridges. Concrete 90. *Highways*, July, **58**, No. 1963, 19–21.
TRINH J. L. (1990). A. S. Nowak (ed.). Structural strengthening by external prestressing. *Bridge evaluation, repair and rehabilitation.* Nato ASI Series E: Applied Sciences, Vol. 187,

ISBN 0 7923 0999 5, 513–523.

TROUT J. F. (1989). Epoxy resin repair of concrete structures. *Proc. Int. Conf. on Structural Faults and Repair*. Kensington, London, 27–29 June.

VASSIE P. R. (1987). *Durability of concrete repairs: the effect of steel cleaning procedures*. TRRL Research Report 109.

WALLBANK E. J. (1989). *The performance of concrete in bridges—a survey of 200 highway bridges*. HMSO, London.

WOOD J. G. M., KING E. S. and LEEK D. S. (1989). Defining the properties of concrete repair materials for effective structural application. *Proc. Int. Conf. on Structural Faults and Repair*. Kensington, London, 27–29 June, 231–235.

5

Protection

5.1 Corrosion countermeasures

5.1.1 Background

In the 1960s, coatings were sometimes applied to concrete bridges to mask surface defects, improve appearance and hopefully to provide additional protection. Some coatings failed after a very short time and, in view of the maintenance implications, the Department of Transport (1966) banned the practice of painting its concrete structures. The Department restated this ban (1975) and still maintains the policy of not painting concrete highway structures for cosmetic purposes.

5.1.2 Waterproofing and drainage

As the corrosion process is fuelled by water and oxygen, the first line of defence is the bridge deck waterproofing layer and the sealing of expansion and articulation joints in the superstructure. Experience shows that these do not remain effectively waterproof for ever and it is good practice to provide falls, channels and outlets so that any leakage that may occur does not pond on the tops of crossheads or cause unsightly staining on other surfaces.

5.1.3 Control of deicing practices

The use of more salt than is necessary in deicing procedures is not only a waste, it adds to the deterioration of bridges. Spreading rates need watching. Non-aggressive alternatives to salt have been tried as deicing agents but they are more expensive and generally less satisfactory than salt. Sadegzadeh and Page (1990) studied the effects of urea on reinforced concrete and found that it does not adversely affect durability except in the unlikely event of an accumulation of urea crystal growth. However, the use of urea poses an environmental pollution problem. Calcium magnesium acetate, CMA, is an alternative which causes less environmental damage.

5.2 Concrete surface treatments

5.2.1 Introduction

Pfeiffer and Scali (1981) presented a comprehensive literature survey and reviewed the various concrete sealing materials available. Twenty-one materials from twenty-one different chemical companies, representative of all of the chemical types most commonly used, were tested for water absorption and chloride intrusion in comparison with a plain uncoated concrete control specimen. The materials tested are listed in Table 5.1 and results are reproduced in Figures 5.1 and 5.2.

Table 5.1. Materials tested

Sealing material	Test numbers
Boiled linseed oil	2
Five epoxies with different amounts of solids	15, 16, 18, 19, 21
Two epoxies containing polysulphide	17, 20
Three materials based on methacrylate	8, 10, 13
Silane	6
Two types of urethane	4, 14
Two materials based on butadiene	7, 11
Chlorinated rubber	5
Silicate	9
Siliconate	3
Siloxane	1
Material based on isobutylene and aluminium stearate	12

After soaking and weighing to establish water and chloride ion transmission, the specimens were air dried and weighed to test water vapour transmission.

Figure 5.2 shows that there is a relationship between salt-water absorption and chloride content; points above the line indicate a chloride screen and points below the line indicate sealers more permeable to chloride ion than to water.

The wide variations between different types and within generic groups could have been partly attributable to different rates of coverage. However, the following five materials, with the details given in Table 5.2, were clearly superior to others.

Table 5.2 Best performing materials

Test no.	Weight gain: %	Loss/ gain: %	Chloride ion content: %	Sealing material type
16	0.29	100	0.008	Epoxy, 50% solids
15	0.57	60	0.044	Epoxy, 100% solids
8	0.61	120	0.047	Methyl methacrylate
4	0.63	107	0.048	Moist-cured urethane
6	0.83	100	0.050	Alkyl-alkoxy silane
—	2.77	66	0.236	Unsealed concrete

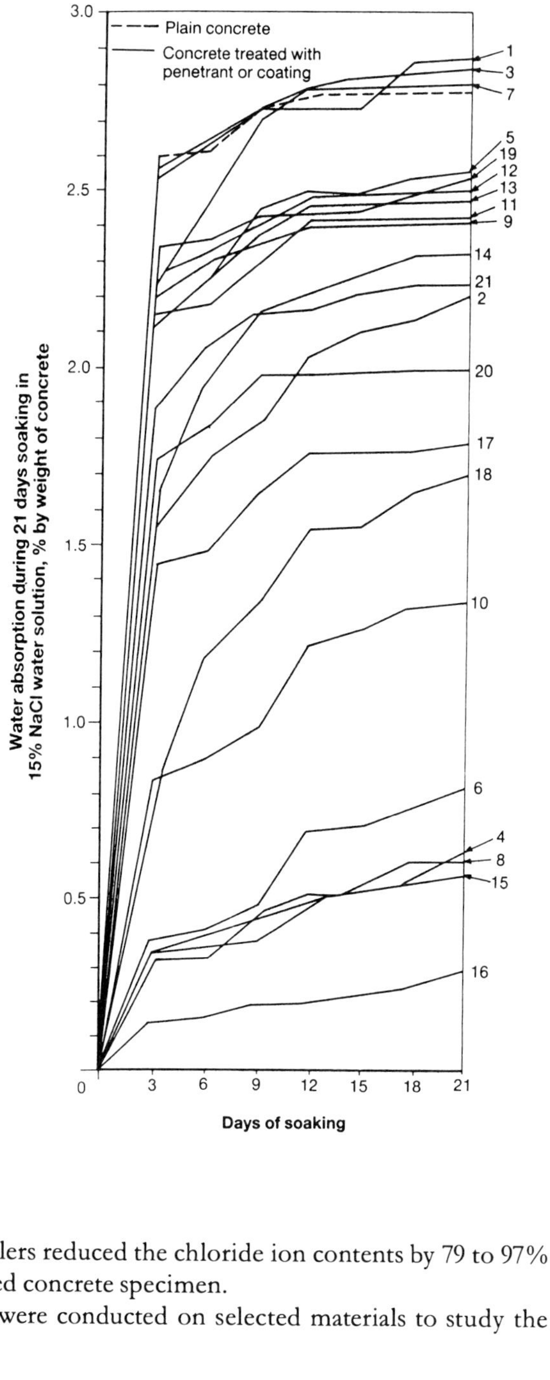

Fig. 5.1. Water absorption characteristics of concrete coated with different sealers (NCHRP report 244, 1981)

These five sealers reduced the chloride ion contents by 79 to 97% compared with the uncoated concrete specimen.

Further tests were conducted on selected materials to study the effects on performance of:

- concrete moisture content at the time of application;
- different coverage rates on chloride intrusion;
- accelerated exposure trials.

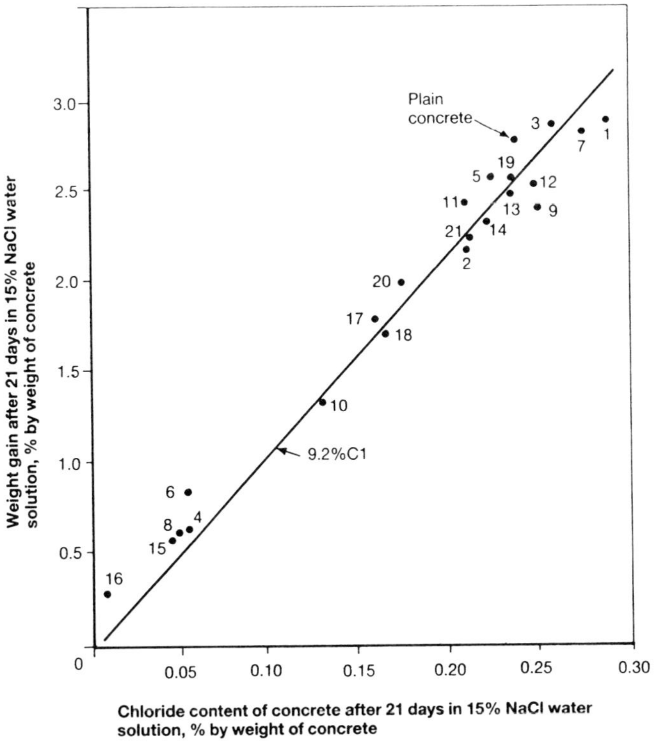

Fig. 5.2. Relationship between water absorption and chloride content of concrete coated with different sealers (NCHRP Report 244, 1981)

Hewlett (1990) made a detailed study of the surface nature of concrete, comprising layers defined by Kreijger (1984) as:

- ☐ cement skin, 0.1 mm thick,
- ☐ mortar skin, 5 mm thick and
- ☐ concrete skin, 30 mm thick.

5.2.2 Classification of treatments

Although some products are designed to protect in more than one way, CIRIA (1987) published a classification on the basis of which Kay (1992) suggested that most materials can be classified in one of the following three main groups:

- ■ coatings/renderings
- ■ pore liners/pore blockers
- ■ sealers.

Kay (1992) illustrates the essential differences between these groups in the degree of penetration of the pore structure of the concrete surface.

CIRIA (1987) suggests sub-division into:

- ☐ thin coatings, up to 1 mm thick,
- ☐ thick coatings, from 1 to 5 mm thick and
- ☐ renderings, thicker than 5 mm.

Coatings, which are applied like paints, and renderings, which are applied by trowel, rely on the formation of a surface barrier to the ingress of liquids and gases.

Pore liners and pore blockers are applied as low viscosity liquids which solidify by solvent evaporation. As pore liners repel water but permit the passage of gases, they can result in long-term drying out of concrete.

In an investigation for TRRL of five types of crystal growth materials marketed as pore blockers, treated reinforced concrete specimens were compared with similar but untreated specimens. The time to corrosion was not delayed and chloride penetration was not reduced by the treatments.

Leeming (1988) reviewed available materials and stressed the importance of careful preparation and application, including environmental conditions. Coatings can be expected to last 10 years and some appear to be performing well after 30 years.

5.2.3 Coating of concrete

It has long been good practice to treat the earth-retaining surfaces of substructures with bituminous or similar coatings. Conventional exterior quality paints are not suitable for use on the exposed surfaces of concrete bridges. A coating which is impervious to moisture will not stay on a concrete surface from which moisture is trying to escape.

Bijen (1989) attributes the binding of coatings either to polyaddition or polymerization reactions between the resin components or to drying by evaporation of the solvents or dispersing agents. He gave the following characteristics of the most important resins.

(*a*) Epoxies—most provide good adhesion to concrete, they show minor polymerization shrinkage, are resistant to light chemical attack and can be formulated to eliminate yellowing almost entirely.
(*b*) Polyurethanes—adhere well to dry concrete, are almost shrinkage-free, are resistant to light chemical attack but not to highly alkaline conditions, also some are susceptible to yellowing.
(*c*) Tar epoxies and tar polyurethanes—more flexible than epoxies and better alkali resistance than polyurethanes but not such good resistance to ultraviolet light, oxidation and temperature changes.
(*d*) Polymethylmethacrylates—very good adherence to concrete, high strength and good resistance to light chemical attack but high polymerization shrinkage.

Drying types include the following.

(i) Acrylics—properties differ for acrylics polymerized in situ or applied in solution or in dispersion but most display good adherence to concrete and good resistance to alkali, oxidation and weathering.
(ii) Chlorinated rubber and vinyl chloride copolymers and other combinations not designed for bridges.

Shaw (1990) reviewed the development of anti-carbonation coatings. These have low permeability to carbon dioxide but adequate permeability to water vapour, which is not very difficult since the water vapour molecule is much smaller than the carbon dioxide molecule. Two coats of a formulation based on copolymers of vinyl acetate or acrylic resin dispersions, with a thickness of 80–120 μm per coat, can give excellent performance on alkaline substrates.

Parker (1990) suggested that the choice of protective coatings should be limited to those with water vapour permeance less than 67 mg/sMN, high chemical and water resistance and good durability. Epoxide, polyurethane epoxide combinations, high build chlorinated rubbers and bitumens and acrylic solutions are given as the more promising materials.

5.2.4 Impregnation of concrete

Bijen (1989) distinguished the following compounds which should remain water-repellent for at least ten years.

- Silicon resins with molecular weight, M > 2000—can only be applied effectively to dry materials with coarser pores than concrete.
- Polymeric alkylalkoxy siloxanes (like resins), M > 1000—very like resins, tend to be sticky and seldom used on concrete.
- Oligomeric alkylalkoxy siloxanes (siloxanes), 600 > M > 400—slightly polymerized silanes in solution with 5 to 10% active content and good impregnation.
- Alkylalkoxy siloxanes (silanes), 200 > M > 100—about 40% silane in low viscosity solvent gives good impregnation and polymerization is catalysed by the concrete but with the disadvantage that the material itself is volatile and could evaporate with the solvent in unfavourable weather conditions.
- Siliconates, 200 > M > 100—water soluble or alcohol soluble with about 40% solids, reaction depends upon carbon dioxide from the air and seldom used on concrete.

Hewlett (1990) noted that silane molecules are similar in size to the water molecule and quoted the following ranking for penetrability

alkylalkoxy siloxanes > silicones > acrylics > epoxies

The Department of Transport (1990) issued guidance and criteria for the impregnation of highway structures using monomeric alkyl (isobutyl)-trialkoxy-silane.

Shaw (1990) stressed the importance of good penetration into the pore structure of the concrete. If the water-repellent silicon resin, resulting from silane/siloxane treatment, is too near the surface it is unlikely to last more than a few years. He also mentioned water-repellent systems based on stearates, but their molecular size is similar to that of the silicon resins and they seldom last more than five years.

Pearson-Kirk (1992), in a general review of the impregnation of concrete, summarized the comprehensive studies by Satoh *et al.* (1989) of five types of alkyl-alkoxy silanes as follows.

- ☐ Penetration depths of silanes were generally 3–5 mm.
- ☐ Pore size volumes and distributions were almost the same for treated as for untreated specimens.
- ☐ Water permeation, evaluated under pressure corresponding to 80 km/hour wind for 48 hours, was less than 5% of that of untreated concrete.
- ☐ Water absorption, after immersion at 20°C for 91 days, was between 20 and 25% of that of untreated concrete.
- ☐ Chloride ion penetration, after immersion in 2.5% NaCl solution at 20°C for 28 days, ranged from 0.1 to 0.6 mm compared with 27 mm for untreated specimens.
- ☐ Accelerated carbonation tests showed that the barrier penetrant treated specimens were not significantly different from untreated specimens with respect to CO_2 diffusion.
- ☐ Accelerated weathering tests showed that the treated specimens had excellent resistance.

Whiting *et al.* (1992) presented an electrical resistance method and a surface absorption method for evaluating penetrating sealers. Both methods are non-destructive, rapid and can be used on horizontal or vertical surfaces.

Pearson-Kirk and Bates (1993) reviewed UK experience of the impregnation of concrete highway structures.

5.2.5 Concrete surface treatment to counter ASR

It was noted (1.3.2) that moisture is one of the essentials for the continuation of alkali-silica reaction; also the importance of providing and maintaining effective bridge deck waterproofing and drainage (1.5.2 and 5.1.2). Various surface treatments to enable concrete to dry out have been studied.

Blight (1991) compared the effectiveness of the following two pore liner penetrants (5.2.4):

- silicone in a hydrocarbon solvent
- alkylalkoxy siloxane

and two coatings (5.2.3):

- cement-based resin-modified slurry
- aqueous emulsion of synthetic resins.

Laboratory tests showed that none of the treatments was completely waterproof. However, the silane was outstandingly better than the others and, after four years' weathering, was still water-repellent and inhibiting carbonation. Parallel field tests on columns showed that, in a semi-arid climate, concrete dries out fairly quickly after surface wetting and does not need to be waterproofed. There was a slight leakage in the seals of drainage ducts which allowed small quantities of water to enter the concrete. Before the surfaces were sealed, this moisture evaporated but the treatments retained this moisture and allowed it to accumulate, an adverse effect!

Miyagawa *et al.* (1991 and 1992) reported surface treatments adopted in the course of maintenance work, stressing the importance of environmental conditions and warning against the possibility of surface treatment even resulting in an adverse effect. They favoured coatings with good crack bridging performance such as urethanes and epoxies; silanes were highly effective but guidance on rates of application was needed. In attempting to control ASR expansion by means of surface treatment, it is necessary to consider permeability not only to the ingress of water but also to the loss of moisture. They examined the long-term performance of reactive specimens treated with:

- epoxy waterproofing
- urethane waterproofing
- silane
- silane with PCM
- silane with MMA
- acrylic
- PET sheet.

Kojima *et al.* (1992), studying the effectiveness of coatings to inhibit ASR, found that the most promising was silane monomer with a polymer cement type of coating.

Swamy and Tanikawa (1992) presented data on the effectiveness of acrylic rubber coatings subjected to outdoor exposure and three accelerated aggressive regimes. A highly elastic acrylic was more effective than an epoxy coating and conserved about 80% of the original concrete strength.

5.3 Coating of reinforcement

5.3.1 Epoxy coated bars

Fusion bonded epoxy coated reinforcement (FBECR) was used increasingly in the USA for more than 20 years. Weyers and Cady (1987) reported favourably on its performance compared with uncoated reinforcement. FBECR was used in the mid 1980s in the parapets of a viaduct at Cardiff and *New Civil Engineer* (1986) reported its use in several new bridges in the UK.

Treadaway, Cox and Davies (1988) found that FBECR was more resistant to chlorides than galvanized or unprotected steel but, at 3.9% chloride concentration by weight of cement, they found underfilm corrosion spreading from damaged areas. Then a notable failure in bridges at Florida Keys reported by Sagues *et al.* (1990) brought the reliability of epoxy coating and the high level of permitted damage very much into question. FBECR is unlikely to find much application in bridge repairs, except possibly in complete redecking or replacing large areas of reinforcement.

5.3.2 Zinc coated bars

Shimada and Nishi (1983) reported the results of five years of marine exposure tests. The corrosion rate for galvanized bars was much slower than hot rolled uncoated bars but the onset of concrete splitting was not delayed significantly.

Treadaway, Cox and Davies (1988) found that galvanizing gave increased protection in carbonated concrete uncontaminated with chloride but, above 0.4% chloride by weight of cement, the risk of corrosion increased and, at high concentrations, approached that of unprotected steel.

In 1988, there were two failures of 50 mm diameter quenched and tempered (QT) post-tensioning tendons which had been galvanized (Glinton) and sherardized (Porth). In each case failure had occurred five days after stressing and was firmly attributed to brittle fracture and not stress corrosion. *New Civil Engineer* (1990) featured the results of the report by McCalls Special Products which found 1.26 ppm of hydrogen in the 200 μm thick galvanizing and 0.6 ppm in the 30 μm thick sherardizing, levels at which the crack initiation stress was greatly reduced. The company had discontinued zinc-coating its products.

Swamy (1990) presented results of tests on resistance to chlorides of galvanized bars in cracked concrete. Less than 10% of bar surface was affected by pitting corrosion after two years of accelerated exposure followed by three years of natural exposure in a tidal environment. In comparison, plain uncoated bars showed extensive corrosion and much deeper pitting but cracking, cover and concrete quality remain the most critical factors in chloride-contaminated concrete.

5.3.3 Priming coat

The Concrete Society (1984) advised that no anti-corrosion treatment is necessary in conjunction with cement-based repair systems. A priming coat, with or without corrosion-inhibiting properties, is usually recommended by suppliers of resin-based systems and should be applied as soon as possible after cleaning the reinforcement.

Parker (1990) stressed the need for priming both the concrete and the steel in epoxy mortar repairs, using the appropriate formulation for wet or dry conditions, applied in accordance with the manufacturer's instructions.

Lambe, Humphrey and Watkins (1990) studied the effectiveness of primers in repair zones and surrounding areas and proposed an electrochemical macrocell test for the evaluation of steel primers.

5.3.4 Insulation of reinforcement

Despite best endeavours to clean the reinforcement (4.3.2), the possibility of chloride pits remaining or developing in the future cannot be excluded. Bijen (1989) noted that this is especially likely to set up new corrosion adjacent to repairs, wherever sufficient oxygen is available. He therefore proposed that reinforcement in repair areas should be first dried, without damaging the concrete by the use of excessive heat, and then coated with an anodic inhibiting solution. He did not consider that the effectiveness of mixed anodic and cathodic inhibitors had been proved and therefore proposed that steel in repair areas should be insulated. Materials to consider, applied in several coats to minimize risk of defects, were:

- solvent-free or low-solvent epoxy
- epoxy with zinc extender
- polymer-modified cement.

Dry concrete is a good electrical insulator but drying may be difficult to achieve as treatment to prevent ingress of water may also prevent water from leaving.

5.4 Cathodic protection

5.4.1 Introduction

The basic mechanism of corrosion is illustrated in Figure 5.3 and that of cathodic protection is illustrated in Figure 5.4.

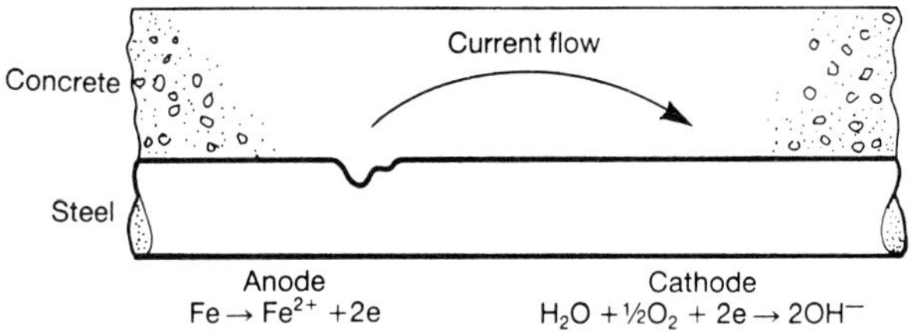

Fig. 5.3. Corrosion cell

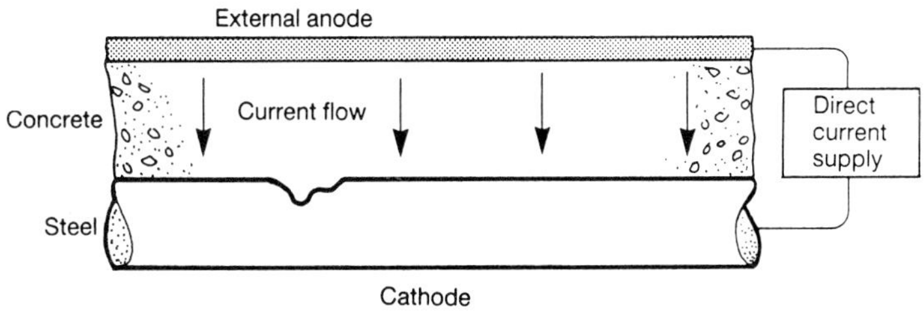

Fig. 5.4. Cathodic protection

Stratfull (1974) reported the experimental cathodic protection of a bridge deck. An electrically conductive asphalt concrete was made by substituting coke breeze for the aggregate and the effectiveness of the system was proved when it stopped the corrosion of steel strips embedded in concrete containing 10% by weight of calcium chloride. The 33 m^2 top reinforcement mat was protected with a total current of 1 A at 1650 mV.

In 1982, following a decade of research and field trials on methods of corrosion control and the repair of corroding reinforced concrete, the US Federal Highway Administration stated its considered view:

> 'The only rehabilitation technique that has proven to stop corrosion in salt-contaminated bridge decks regardless of the chloride content is cathodic protection.'

Vermani (1983) reported progress on cathodic protection of bridge decks and assessed different anodes as follows.

- Slotted anodes were economical but currents should be kept to the minimum to stop corrosion. Hydrochloric acid attacked the concrete in some slots although this problem lessened as the amount of chloride ion present decreased.
- Strips covered with rigid overlay, the larger area of which minimized the acidity problem. The strip area was about 8% of the concrete surface area.
- Conductive overlays appeared to be ideal since the ionic reaction is distributed over the whole area, although chlorine gas evolved had produced some delamination.

Cathodic protection was also being applied to substructures trying a variety of anode systems.

Starting with a literature review, Kendall and Lewis (1984) reported on the feasibility of cathodic protection for bridge decks. Reinforcement is protected by making its electrical potential sufficiently negative to override any incipient corrosion cells. To do this, a suitable conducting element is installed and connected to the positive side of a power supply to form an anode. The negative side is connected to the steel reinforcement which becomes cathodic and the impressed current (electrons) reduces oxygen with the moisture in the concrete acting as an electrolyte. Depending on the particular concrete characteristics and chloride concentrations, the potential of the reinforcement has to be between −650 and −1135 mV (CSE), values between −700 and −800 mV (CSE) being most frequently quoted for minimum protection. Fromm (1977) obtained polarized potentials between −850 and −1100 mV (CSE) on bridge deck reinforcement by applying a potential of −1300 mV (CSE) at the anode.

It should be remembered that waterproofing of bridge decks was not an established practice in the USA. Cathodic protection could only be applied to a waterproofed bridge deck if the waterproofing layer is taken up before installing the anode; otherwise it would act as an insulator.

Because concrete is so heterogeneous, some overprotection may be required to ensure that no areas fall below the prescribed level of protection. Unfortunately this has the disadvantage that overprotecting the steel tends to produce hydrogen which may reduce bond strength. Concern about hydrogen embrittlement relates to highly stressed medium carbon steels used for prestressing and not to ordinary reinforcement. This problem would need to be considered in applying cathodic protection to pretensioned tendons in precast concrete members but CP is not applicable to post-tensioning tendons in contact with metal ducts which would interfere with the protecting currents.

Better Roads (1985) published an article describing cathodic protection and the problem of finding a stable anode capable of distributing the impressed current uniformly over a large area. The following systems were briefly described:

- a grid of closely spaced anodes;
- less closely spaced anodes within a conductive material;

- ☐ flexible grid of copper conductors coated with conductive polmer and overlaid with cast or sprayed concrete;
- ☐ carbon fibre coated with polymer resin and overlaid with latex-modified concrete.

Hope and Ip (1987) found that cathodic protection of the mild steel reinforcement in a thick post-tensioned deck slab would not adversely affect the prestressing steel; nor would it in a thin slab deck, provided that the current density on the surface did not exceed 20 mA/m^2. This study recommended caution in considering cathodic protection of the prestressing steel in pre-tensioned concrete beams.

Boam (1988), discussing options for repair, pointed out the advantages of cathodic protection compared with alternative procedures. Provided there is no loss of structural integrity, only repairs to spalling and delamination of the concrete are required. There is no need to remove large volumes of concrete because it is contaminated with chloride. Cathodic protection will prevent future corrosion including incipient corrosion at the boundaries of repaired areas. More needs to be known about the life of anode systems to enable bridge authorities and engineers to evaluate whole life costs with more confidence.

Wyatt (1988) reviewed cathodic protection as the long-term repair strategy. In addition to conductive overlays, slotted anodes and distributed mesh anodes, used for bridge decks and car park decks, very thin conductive coatings, about 400 μm, have been developed for application to vertical surfaces and soffits.

The Concrete Society (1989) published the Report of a joint Working Party with the Corrosion Engineering Association. The information and guidance is concentrated on the cathodic protection of reinforced concrete structures above ground.

Apart from sacrificial anodes, which are uncontrolled and not used on bridges, the following options are available:

- deck anodes
  - — conductive asphalts
  - — slotted systems
  - — conductive mesh systems
- substructure anodes
  - — conductive mesh systems
  - — conductive coatings
  - — discrete anodes
  - — sprayed metal coatings.

Manning (1990) presented the advantages and disadvantages of three generic types of anode systems:

- continuous surface anodes
- discrete anodes
- mesh anodes.

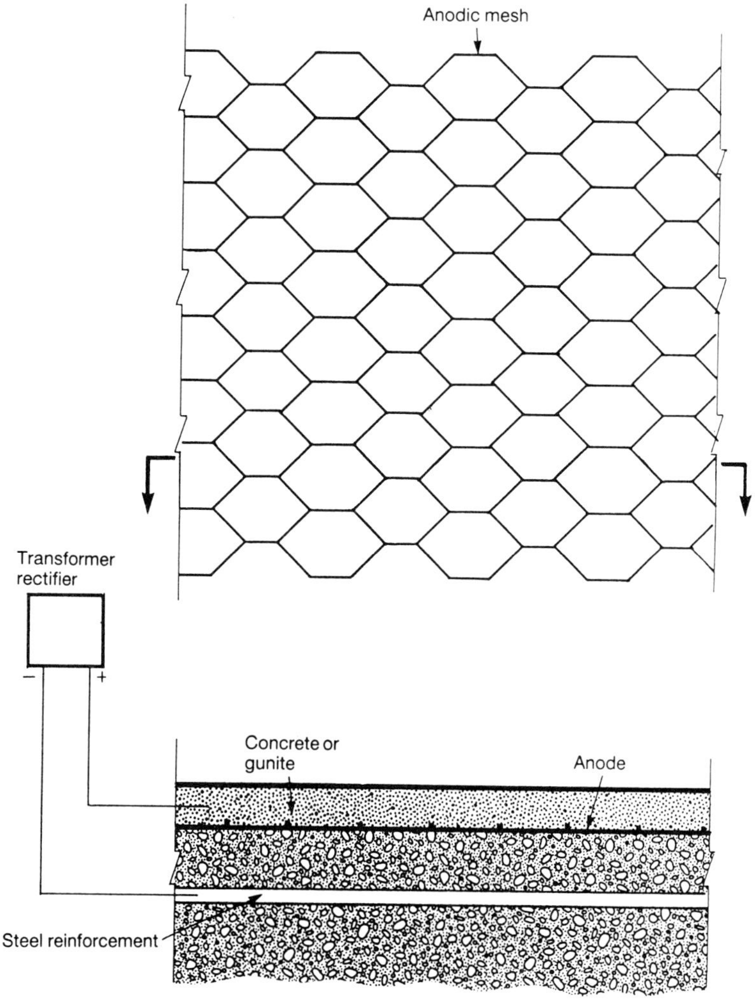

Fig. 5.5. Cathodic protection with anodic mesh (PIARC, 1991)

The conductive asphalt system used in Ontario had performed well and cathodic protection is also viable for substructures.

The PIARC Technical Committee on Bridges (1991) reviewed the cathodic protection of bridge decks in some detail, including principles of operation and the evolution of anodic systems as follows:

(*a*) simple anodes in conductive layer, overlay,
(*b*) slotted anodes in deck concrete, no overlay,
(*c*) anodes in conductive mortar surfacing,
(*d*) anodic net covered with mortar, overlay (Figure 5.5),
(*e*) anodic grill with mortar, overlay.

5.4.2 Monitoring

The following main contenders can be considered for the routine monitoring of cathodic protection installations:

- potential criteria
- E log I testing
- potential decay.

Gurusamy and Gheoghegan (1990) reported the suitability of durable embeddable silver/silver chloride reference electrodes to monitor cathodic protection systems. They were manufactured in the laboratory and calibrated against a standard calomel electrode to a potential of -44.5 ± 4.5 mV with a rejection rate of 25%.

Vassie (1991), in a comprehensive guide to the half-cell potential technique, gave potentials of different reference electrodes by comparison with the absolute standard hydrogen electrode (SHE). Values are shown in Table 5.3.

Table 5.3. Standard half-cell potentials

Reference electrode	Potential against SHE at 25°C
Saturated copper/copper sulphate	316 mV
Saturated calomel	242 mV
Silver/silver chloride	200 mV

McKenzie (1991) monitored the performance of cathodically protected reinforced concrete specimens with chloride contamination compared with similar but unprotected specimens. Cathodic protection significantly reduced both the extent and severity of corrosion over five years with no detectable effect on the steel–concrete bond strength.

5.5 Desalination and realkalization

Manson *et al.*, Weyers *et al.* (1983) reported preliminary experiments which confirmed earlier reports of the successful removal of salt from concrete by electrochemical means. In one well-documented study by Morrison *et al.* (1975) salt concentrations of 6 to 12 kg/m^3 were removed in this way in times up to 100 hours but the process has the effect of increasing the permeability of the concrete fivefold. To offset this the treated concrete could be polymer impregnated.

Reporting a feasibility study within the Strategic Highway Research Program, Asaro, Gaynor and Hettiarachchi (1990) presented a system for electrochemical chloride removal by injecting an inhibitor at a typical rate of 95 μg/cm^2 per hour with an electric field of 5 to 10 V/cm and current density about 1 mA/cm^2. Equipment similar to that used for cathodic protection is only required to be installed for a few days.

Anderson (1990) described two techniques which involve passing a current

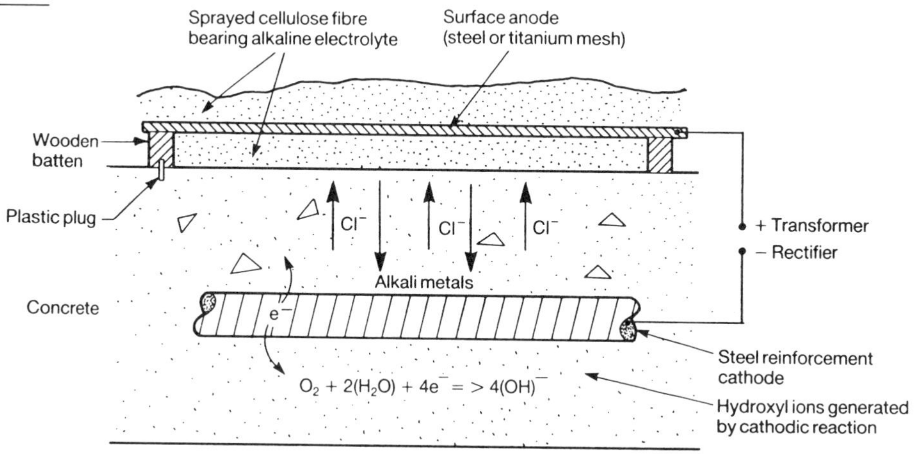

Fig. 5.6. Chloride extraction and realkalization (Anderson, 1990)

through the concrete using an external anode mesh attached to the concrete surface as shown in Figure 5.6.

In desalination, chloride ions migrate to the surface anode and are removed. In realkalization, the impressed current forces alkali ions, particularly sodium, to the cathodic reinforcement where the reaction converts water and oxygen into hydroxyl ions and repassivates the steel.

Desalination takes typically from three to eight weeks and may result in repassivation of the reinforcement by generation of hydroxyl ions. If further realkalization is required this can then usually be completed in three to six days.

Research Focus (1992) reported that the Building Research Establishment has started to investigate the technique. Core samples will be taken from a building being realkalized and the effectiveness of the process will be monitored leading to advice in 1995. Realkalization and the withdrawal of chlorides by electro-osmosis was described by Demars (1993).

The Strategic Highway Research Program has a project in progress to develop a cost-effective chloride removal system. It has been noted that all of the chloride is not removed and will eventually remigrate back to the reinforcing steel if preventive measures are not taken. Results are awaited.

5.6 References

Anderson G. (1990). Chloride extraction and realkalization of concrete. *Hong Kong Contractor*, July–Aug., 19–25.

Asaro M. F., Gaynor A. T. and Hettiarachchi S. (1990). *Electrochemical chloride removal and protection of concrete bridge components (injection of synergistic corrosion inhibitors)*. SHRP-S/FR-90-002. National Research Council, Washington DC, USA.

Better Roads (1984). Fighting bridge decay from within. *Better Roads*, Nov., 32–34.

Bijen J. M. J. M. (1989). Maintenance and repair of concrete structures. *Heron*, **34**, No. 2, 49–63.

Blight G. E. (1991). A study of four waterproofing systems for concrete. *Mag. Concr. Res.*, **43**, No. 156, 197–203.

Boam K. (1988). Concrete highway structures—current experience of investigation and repair. *Concrete Repair Conference*. London Tara Hotel, 16 Mar., Palladian Publications Ltd.

Calder A. J. J. and Thompson D. M. (1988). *Repair of cracked reinforced concrete: assessment of corrosion protection*. TRRL Research Report 150.

Ciria (1987). *Protection of reinforced concrete by surface treatments*. Technical Note 130. Construction Industry Research and Information Association, London.

Concrete Society (1984). *Repair of concrete damaged by reinforcement corrosion*. Technical Report No. 26.

Concrete Society (1989). *Cathodic protection of reinforced concrete: Report of a joint Concrete Society and Corrosion Engineering Association Working Party*. Technical Report No. 36.

Demars P. (1993). Re-alkalisation and withdrawal of chlorides in concrete by means of electro-osmosis. *Proc. 2nd Int. Conf. on Bridge Management*. University of Surrey, Guildford, 18–21 Apr.

Department of Transport (1966). *Colour and texture in bridge design*. Technical Memorandum BE 2.

Department of Transport (1975). *Painting of concrete highway structures*. Technical Memorandum (Bridges) BE8/75.

Department of Transport (1990). *Impregnation of highway structures*. Departmental Advice Note BA33/90.

Gurusamy K. N. and Gheoghegan M. P. (1990). The long term performance of embedded reference electrodes for cathodic protection and in situ monitoring of steel in concrete. *Proc. 3rd Int. Conf. on Corrosion of Reinforcement in Concrete Construction*. Wishaw, Warwickshire, 21–24 May. Published for the Society of Chemical Industry by Elsevier Applied Science, 333–347.

Fromm H. J. (1977). Cathodic protection of rebar in concrete bridge decks. *Materials Performance*, Nov.

Hammersley G. P., Dill M. J. and Darby J. J. (1990). Investigation into the effectiveness of silane for reducing corrosion activity in a chloride-contaminated reinforced concrete bridge structure. *Bridge management—inspection, maintenance, assessment and repair*. Elsevier Applied Science, London and New York.

Hewlett P. C. (1990). R. Dhir and J. W. Green (eds). Methods of protecting concrete—coatings and linings. *Proc. Int. Conf. on Protection of Concrete*. 11–13 Sept., University of Dundee, 105–134.

Hope B. B. and Ip A. K. C. (1987). *Application of cathodic protection to prestressed concrete bridges*. Report ME-87-01, Ontario Ministry of Transportation, USA.

Kay T. (1992). *Assessment and renovation of concrete structures*. Longman Scientific and Technical, Essex.

Kendall K. and Lewis D. A. (1984). *Bridge decks: cathodic protection*. TRRL Contractor Report 4.

Kojima T., Miyagawa T., Nakano K., Yamaguchi Y. and Kobayashi K. (1992). Effect of coating to inhibit AAR in concrete structures. *Proc. 9th Int. Conf. on AAR in Concrete*. Queen Elizabeth II International Conference Centre, London.

Kreijger P. C. (1984). The skin of concrete, composition and properties. *Materiaux et Constructions*, **17**, No. 100, 275–283.

Lambe R. W., Humphrey M. J. and Watkins T. F. (1990). Development of materials for repair and protection of reinforced concrete. *Proc. 3rd Int. Conf. on Corrosion of Reinforcement in Concrete Construction*. Wishaw, Warwickshire, 21–24 May. Published for the Society of Chemical Industry by Elsevier Applied Science, 471–485.

Leeming M. (1988). Surface coatings for concrete. *Concrete Repair Conference*. London Tara Hotel, 16 Mar., Palladian Publications Ltd.

McKENZIE M. (1991). *Cathodic protection of reinforced concrete: effectiveness on initially uncorroded steel*. TRRL Research Report 328.

MANNING D. G. (1990). Cathodic protection of concrete highway bridges. *Proc. 3rd Int. Conf. on Corrosion of Reinforcement in Concrete Construction*. Wishaw, Warwickshire, 21–24 May. Published for the Society of Chemical Industry by Elsevier Applied Science, 486–497.

MANSON J. A., SLUTTER R. G. and VANDERHOFF J.W. (Lehigh University) and WEYERS R. E., CADY P. D., KLINE D. E., MOORE G. R., WILSON D. T. and WONG K. H. (Pennsylvania State University) (1983). *Long-term rehabilitation of salt-contaminated bridge decks*. Transportation Research Board, National Cooperative Highway Research Program Report 257.

MIYAGAWA T., HISADA M., INOUE S. and FUJII M. (1991). Effect of concrete surface treatment on expansion due to alkali-silica reaction. *Concrete Library of JSCE*, Dec., No. 18, 237–261.

MIYAGAWA T., INOUE S., FUJII M. and UEHATA N. (1992). Effect of concrete surface treatment on expansion due to alkali-silica reaction. *Proc. 9th Int. Conf. on AAR in Concrete*. Queen Elizabeth II International Conference Centre, London.

MORRISON G. L., VIRMANI Y. P., STRATTON F. W. and GILLILAND W. J. (1975). *Chloride removal and monomer impregnation of bridge deck concrete by electro-osmosis*. Report No. FHWA-Ks-RD-74-1.

NEW CIVIL ENGINEER (1986). Brian Dumbleton. Coated bars bid to cut corrosion. *New Civil Engineer*, 21/28 Aug. 1986.

NEW CIVIL ENGINEER (1990). Stressing success. *New Civil Engineer*, 26 Apr. 1990, 18.

PARKER A. J. (1990). Technical Report. *Reinforced concrete: structural considerations and choice of products for repairs*. Federation of Epoxy Resin Formulators and Applicators, 2nd edn.

PEARSON-KIRK D. (1992). Impregnation of concrete. *Durability of Highway Structures Symposium*. Institution of Highways and Transportation, Taunton, 12 Nov.

PEARSON-KIRK D. and BATES G. J. (1993). Impregnation of concrete highway structures. *Proc. 2nd Int. Conf. on Bridge Management*. University of Surrey, Guildford, 18–21 Apr.

PFEIFFER D.W. and SCALI M. J. (1981). *Concrete sealers for protection of bridge structures*. National Cooperative Highway Research Program Report 244. Transportation Research Board, Washington DC, USA.

PIARC TECHNICAL COMMITTEE ON ROAD BRIDGES (1991). *Cathodic protection of bridge decks*. Drafted by Paolo Baldo. Permanent International Association of Road Congresses.

RESEARCH FOCUS (1992). *Repairing reinforced concrete*. No. 12, Jan., 4.

ROMAGNOLO M., FERRARI S. and GRANDI M. (1990). A. S. Nowak (ed.). Cathodic protection of reinforced concrete structures—Experimental system installed on a viaduct in service on the IRI-ITALSTAT network. *Bridge evaluation, repair and rehabilitation*. Nato ASI Series E: Applied Sciences, Vol. 187. ISBN 0 7923 0999 5.

SADEGZADEH M. and PAGE C. L. (1990). *Effects of urea on the durability of reinforced concrete*. TRRL Contractor Report 218.

SAGUES A., POWERS R. and ZAYED A. (1990). Marine environment corrosion of epoxy-coated reinforcing steel. *Proc. 3rd Int. Conf. on Corrosion of Reinforcement in Concrete Construction*. Wishaw, Warwickshire, 21–24 May. Published for the Society of Chemical Industry by Elsevier Applied Science, 539–549.

SATOH Y., OHAMA Y., DEMURA K. and TACHIBANA K. (1989). Improvements in barrier properties of concrete by silane impregnation. *Proc. 3rd Int. Conf. on Deterioration and Repair of Reinforced Concrete in the Arabian Gulf*. Bahrain Society of Engineers, Oct., 689–699.

SHAW J. D. N. (1990). The development of coatings for concrete materials and case histories. *Proc. Int. Conf. on Protection of Concrete*. Dundee, 11–13 Sept.

SHIMADA H. and NISHI S. (1983). Sea water corrosion attack on the concrete blocks embedding zinc galvanized steel rebars. Paper presented at meeting: *Corrosion of reinforcement in concrete construction*. London, 13–15 June, Society of Chemical Industry, 407–418.

STRATFULL R. F. (1974). *Experimental cathodic protection of a bridge deck*. Transportation Research Record No. 500, 1–15.

SWAMY R. N. (1990). Resistance to chlorides of galvanized rebars. *Proc. 3rd Int. Conf. on Corrosion of Reinforcement in Concrete Construction*. Wishaw, Warwickshire, 21–24 May. Published for the Society of Chemical Industry by Elsevier Applied Science, 586–600.

SWAMY R. N. and TANIKAWA S. (1992). Acrylic rubber coating to control alkali aggregate reactivity. *Proc. 9th Int. Conf. on AAR in Concrete*. Queen Elizabeth II International Conference Centre, London.

TREADAWAY K. W. J., COX R. N. and DAVIES H. (1988). *Corrosion-protected and corrosion-resistant reinforcement in concrete*. Building Research Establishment Information Paper IP 14/88.

VASSIE P. R. (1991). *The half-cell potential method of locating corroding reinforcement in concrete structures*. TRRL Application Guide 9.

VERMANI Y. P. (1983). FCP Annual Progress Report, 2–55.

WEYERS R. and CADY P. (1987). Deterioration of concrete bridge decks from corrosion of reinforcing steel. *Concrete International*. Jan., **9**, No. 15.

WHITING D., OST B., NAGI M. and CADY P. D. (1992). Vol. 5: methods of evaluating the effectiveness of penetrating sealers. *Condition evaluation of concrete bridges relative to reinforcement corrosion*. SHRP-S/FR-92-107.

WONG K. H., WEYERS R. E. and CADY P. D. (1983). The retardation of reinforcing steel corrosion by alkyl-alkoxy silane. *Cem. Concr. Res.*, **13**, 778–788. Pergamon Press.

WYATT B. S. (1988). Cathodic protection of reinforced concrete—the long-term repair strategy. *Concrete Repair Conference*. London Tara Hotel, 16 Mar., Palladian Publications Ltd.

6

Surveys, studies and trials

6.1 Surveys

6.1.1 Swiss bridge maintenance—*New Civil Engineer* (1983)

In the Swiss canton of Graubünden it was the practice to salt some roads and not others; the difference in the damage to concrete was marked. With typical bridge repairs costing half as much as new bridges, the policy was to give priority to repair and protection of existing bridges. Many of the canton's stock of 1500 bridges were built before 1973, when bridge deck waterproofing was introduced, some three years after winter salting had started and its damaging effect was recognized. Bridges built by that pioneer of reinforced concrete, Robert Maillart, include those at Zuoz (2.2.6) and Salginatobel.

The Salginatobel bridge spans 590 m. Black paint on the north side of the arch slab provided better protection against salt than the white paint on the south side.

The general approach to repair of reinforced concrete was to hack off defective cover, sandblast the bars and replace the cover with polymer-modified cement mix which is more akin to the parent concrete than epoxy resin mixes. Overslabbing was done to make short span simply supported viaducts continuous with fewer joints through which salt laden water could drain.

6.1.2 New Zealand bridges—Rowe, Freitag and Stewart (1984) and Booth, Sweetman and Wolfe (1987)

Rowe, Freitag and Stewart (1984) reported the examination of concrete quality in 320 bridges in the Gisborne, Hawke's Bay, Wairarapa, Manawatu, Taranaki and Wellington regions showing that:

- ☐ climate influences plastic and drying shrinkage
- ☐ climate influences reinforcement corrosion
- ☐ corrosion spalling was the most destructive defect found
- ☐ design details, mix design and site practices decided the ultimate durability of the structure and
- ☐ failure of repairs was usually due to inadequate preparation.

Booth, Sweetman and Wolfe (1987) reported a survey of 630 bridges, representing about 5% of the national stock, which were examined in 1983 to establish the extent of maintenance problems and identify areas for research to improve bridge performance. In a study of the effectiveness of repairs, problems highlighted included expansion joints and repair of spalled or cracked concrete.

6.1.3 German experience—Lebek (1985) and Goehler (1990)

Lebek (1985) reported a survey of problems related to concrete bridge deterioration and maintenance, starting with background data on bridge construction from 1965 to 1984. Well designed and carefully constructed bridges showed less than average deterioration. Defects included cracking, spalling, scaling, accident damage and chloride-contamination which had become a major problem. Wherever possible, the root causes should be corrected before repair is undertaken. Ideally all chloride-contaminated concrete should be removed but this is not always possible.

Goehler (1990) reported on four major bridges built between 1950 and 1953 with thin webs and slabs and small construction depths. The main defects were due to uncompacted concrete and carbonation, ungrouted tendons and defective waterproofing. The performance of these bridges was compared with that of others of similar design; rehabilitation is described.

6.1.4 North American experience—Hay and Virmani (1985), Lacroix (1985), Bickley *et al.* (1986), Gordon (1989) and Harm (1993)

Hay and Virmani (1985) reported that bridge deck deterioration had been a problem in the United States since the early 1950s and a critical problem since 1960. The major cause of deterioration was identified as corrosion of the reinforcing steel due to the increased use of deicing salts since the late 1960s. Epoxy-coated reinforcing steel was the most widely used protective system to control the problem in new construction. Corrosion of reinforcing steel in pier caps, beams and pier columns had been a problem in recent years. While epoxy-coated reinforcement has been used, a less costly protective system was being sought for these members.

The major emphasis of current research programmes in the United States was on rehabilitation of bridge decks and other bridge members. The only method believed to arrest corrosion of reinforcing steel was cathodic protection. The first practical application was on the deck of Sly Park Bridge, California. This, and other applications of the same system, had performed satisfactorily but were too costly for general use and the search continued for more economical and reliable cathodic protection systems and for methods which removed chloride ions by impressing a direct current on the reinforcing steel.

Lacroix (1985) reported a bridge deck survey in Illinois. Performance was monitored from 1977 through 1984 on 20 mainline interstate bridges with chloride-contaminated decks, rehabilitated by removal and replacement of only unsound concrete and then waterproofed and overlaid. All decks had performed well and none needed immediate rehabilitation by the end of the study. He concluded that patching and then sealing the surface of chloride-contaminated but sound concrete could be a cost-effective alternative to total deck replacement. Harm (1993) had no updated information.

Bickley *et al.* (1986) reported information from Canada and the USA on methods used to detect deficiencies and to repair concrete bridge superstructures and substructures. Many of the respondents did not yet perceive any significant problem with substructures. One interesting observation was that some materials performed well in some locations but not in others, which suggested that methods and workmanship were important rather than choice of material.

Visual and sounding inspections were universal. Coring was considered essential when preparing for structural repairs and many States take cores to determine chloride concentrations. Access difficulties preclude coring as a routine technique. Half-cell potentials (CSE) were generally precluded by access problems and were not used extensively by most agencies.

Ultrasonic pulse velocity measurements were used with varying success. Covermeters were used but not very widely. Thermography was very much in the experimental stage but had been successful in detecting water in hollow box beams.

Removal and replacement was practised by most agencies, the majority using Portland cement concrete but epoxy or latex-modified concrete was also used. Patch repairs met with varying success; failures were usually due to insufficient preparation of the substrate. Dowels were used in some States. Epoxy was used frequently as a bonding agent and by some to coat reinforcement. There was general agreement that removal should be to sound concrete but the development of criteria for removing concrete was seen as a major need. In Illinois, when concrete was removed to behind reinforcing steel, calcium nitrite was used as an admixture to improve protection against corrosion. Epoxy injection of cracks was practised in many States. Shotcrete had found acceptance with many organizations and many States required pre-qualification of operatives. Latex-modified shotcrete had been used successfully in Canada but its rough surface finish was not always acceptable. Cathodic protection had been applied to substructures at three locations. Most agencies seal repairs for several reasons:

- to reduce chloride intrusion;
- for aesthetic reasons;
- to protect the repair.

Some authorities leave the top surfaces unsealed to allow moisture to escape.

Gordon (1989) cited a recent FHA finding that a highway bridge designed for durability would become functionally obsolete before it became structurally deficient. The average age for replacement was 70 years and during this lifetime the bridge would have had one rehabilitation. He presented the following statistics pertaining to a total of 577 710 highway bridges in the USA:

Functionally obsolete	18%
Structurally deficient	24%
Total number deficient	42%

6.1.5 Swedish bridges—Ingvarsson and Westerberg (1985) and Ingvarsson (1988 and 1990)

In a State-of-the-Art Report, Ingvarsson and Westerberg (1985) considered that there was extensive knowledge of durability, maintenance and repair of bridges, including research into environmental factors and deterioration mechanisms. However, the main problem was making the knowledge accessible through manuals, information and education. They pointed to a need for further research on:

- deterioration due to environmental conditions
- interaction between environmental conditions and other factors such as cracking, fatigue, overloading.

The latter is required for whole life assessment of alternative repair proposals.

Ingvarsson (1988) gave the following most common forms of deterioration. Freeze-thaw, chloride and sulphate attack, alkali-silica reaction and corrosion of reinforcement for which the remedies are efficient drainage and waterproofing, surface impregnation with silanes, the use of rust-proof reinforcement and cathodic protection. Details were given of Swedish requirements for repair and strengthening and the factors dictating the choice between repair and renewal.

Ingvarsson (1990) gave the numbers of road bridges shown in Table 6.1.

Table 6.1. Numbers of Swedish road bridges

Patrimony	Stockholm	Gothenburg	Others	All Sweden
State	0	0	10800	10800
State-municipal	520	340	940	1800
Municipal	240	240	2220	2700
Total	760	580	13960	15300

Concrete bridges account for some 80% of the bridge stock for which the Swedish National Road Administration is responsible. Of these 9000 bridges, 400 were prestressed. Seven thousand were built before 1964, when new material requirements, including air entrainment, were introduced. Ongoing inspections suggested that 30% had shortcomings in waterproofing and that 10% were damaged by salt and frost. Only one-third of the 2700 municipal stock had been built before 1964. The Swedish BMS (1.1.2 and 7.2.2) used the following condition classes:

- (0) undamaged on the occasion of the inspection
- (1) defective within ten years

- (2) defective within three years
- (3) defective on the occasion of the inspection.

Average condition classes for the bridges were as follows.

- ☐ European and national roads (0.38)
- ☐ Primary county roads (0.48)
- ☐ Secondary county roads (0.46)
- ☐ Tertiary county roads (0.44)
- ☐ Entire road network (0.44)

6.1.6 Nineteen Scottish bridges—Vassie (1986 and 1987) and Brown and Cochrane (1989)

Twelve concrete bridges, of various types, were constructed in the early 1930s on the A82 trunk road between Orchy and Invergarry in Scotland. They experience a harsh environment and have been subjected to deicing salt since the mid 1960s which resulted in varying degrees of cracking, spalling, leakage and frost damage.

The superstructures were surveyed for reinforcement corrosion using half-cell potential, resistivity, carbonation, chloride and cover measurements and the results have been reported by Vassie (1986). The quality of the concrete was variable but quite good, considering practices in the 1930s. In the decks, chloride concentrations typically exceeded 0.5% by weight of cement with carbonation depths less than 10 mm. On the beams and soffits, corrosion was due to the synergistic effect of carbonation (typically 5–30 mm deep) and chlorides (with mean concentration about 0.2% by weight of cement). Depth of cover, carbonation and chloride measurements suggested that the reinforcement could be expected to corrode. This was supported by the half-cell potential measurements and visual evidence on beams and soffits of four out of eight bridges.

Vassie emphasized the danger of drawing conclusions from one or two tests alone and reported (1987) in further detail on chloride concentrations and resistivity observations on eight of the bridges.

Brown and Cochrane (1989) reported on bridges on the Renfrew Bypass, part of the M8 south-west of Glasgow, constructed in the late 1960s. Seven of the bridges had in-situ reinforced concrete continuous decks, the intermediate supports being precast columns with Maillart hinges top and bottom. During routine inspection some staining and spalling was found on all bridges and it was noted that at four bridges several bottom hinges showed evidence of corrosion while the others were apparently sound.

A full corrosion survey was commissioned in 1987. The visual survey revealed corrosion of the helical bursting steel which was causing spalling around the throats of the four hinges. This appeared to have originated at poorly plugged holes used in the erection procedure. In addition, construction joints in the deck slabs and road joints at abutments were leaking and there was rust staining between the column heads. The cover meter survey found 20 to 40 mm generally except for 10% of readings which were less than

20 mm on six bridges and 41% on one bridge. The majority of the half-cell potentials were numerically less than −200 mV and most resistivity readings were greater than 12 000 Ωcm. This suggested little probability of corrosion, except near leaking deck joints and towards the bases of the columns. Carbonation depths were from 0 to 6 mm. It was established that 0.31% chloride ion was present from the use of calcium chloride in the construction of the columns and chloride concentrations in the 0 to 25 mm cover ranged from 0.03 to 1.07 with an average of 0.6% chloride ion by weight of cement. Concentrations at soffits were generally less than 0.2% except near leaking joints where they reached 0.6 to 1.07%. In abutments the mean maximum was 0.83 and the absolute maximum 2.71%.

Remedial actions included waterproofing joints, repair of spalled areas with proprietary cementitious material and the application of surface coatings where cover was deficient. Rehabilitation of the hinges presented the major problem. Obeche wood packings in the hinges were replaced with a non-fibrous material and some of the lower hinges were replaced with mechanical hinges.

6.1.7 Performance of 92 British bridges—Brown (1986)

The results of the study by Brown (1986) were considered under the following broad headings.

(*a*) Comparing old with new bridges.
(*b*) Relationship between concrete quality and durability.
(*c*) Other relationships between parameters.

The first comparison was noted in 1.5.1. Generally, modern bridges have better concrete cover than older bridges and suffer mainly from chloride attack.

The second part of the study was based on the general durability equation

$$T \propto \mathrm{f}_1(R)\mathrm{f}_2(Q)\mathrm{f}_3(E)$$

in which, for one particular area of concrete,

T is the age of the concrete when spalling occurs,
R is the cover,
Q is the quality of the concrete and
E is a measure of the severity of the environment.

This led to the relationship

$$\sqrt{T} = 2 \times 10^{-3} RC/\mathrm{Cl}_0$$

in which

T is age in years at spalling,
C is cement content in kg/m^3 and
Cl_0 is the extrapolated chloride concentration at the concrete surface.

Critical chloride concentrations were then expressed as in Table 6.2.

Table 6.2. Effective chloride concentrations

Location	Concentrations: kg/m^3
At leaks	7·5
Marine exposure	10
On new bridges without leaks and without traffic spray	2
On old bridges without leaks or traffic spray	1

Following the same argument, durability equations were set up for capillary porosity, gas permeability and water absorption as measures of concrete quality. Sound and spalled areas were separable by straight lines but with neither more nor less confidence than that based on water/cement or cement content.

The study finished on a cautionary note. The scatter of the results reflected the variability of concrete in the field, which should be borne in mind when attempting to apply laboratory results to the real world. This applies especially to durability which often depends on extreme values rather than the average quality of the material.

6.1.8 Concrete bridges in Belgium—Van Begin (1987)

The Belgian Public Works Ministry manages about 5200 large bridges. About 2080 are of reinforced concrete and 1560 are of prestressed concrete construction. The average bridge has an area of 1200 m^2, 22 m spans and is 22 years old. After seven years of systematic inspection, shortcomings were assessed as shown in Table 6.3.

Table 6.3. Incidence of shortcomings

Type of shortcoming	% of cases
Geometrical faults, mainly in cantilever bridges	2 – 4
Cracking of concrete	30 – 80
Corrosion of reinforcing and prestressing steel	20 – 80

All bridges were provided with waterproofing but leakage of deicing salt remained the most serious aggressive factor.

6.1.9 Random 200 English trunk road bridges—Wallbank (1989)

Many of the Department of Transport's 6000 concrete bridges had been closely investigated for particular problems of deterioration but information so gathered did not provide an overview of the condition of the whole stock of bridges. This was obtained in this study by the analysis of data from a sample of 200 bridges selected at random but checked to be representative. For instance, the age distribution of the sample followed that of the total stock of bridges with 82% built in the 1960s and 1970s. Bridges in good, fair and poor condition were all subjected to the same visual inspection and sampling and testing of selected areas of concrete for:

- ☐ half-cell potential
- ☐ chloride content
- ☐ sulphate content
- ☐ carbonation depths
- ☐ minimum cover
- ☐ petrographic examination.

Minimum cover ranged from 0 to 130 mm with 2% less than 20 mm and 5% greater than 80 mm. Mean cover was not related to age.

Carbonation depths were very variable in bridges of all ages. Very few had depths of 20 mm or more and 90% had depths less than 5 mm. Mean depths were slightly less on soffits than on supports and generally only slightly greater in older bridges. The deepest, 35 mm, was found in honeycombed concrete.

Of the 200 bridges, 144 had chloride ion content 0.2% or more by weight of cement in the outer 50 mm of concrete, 31% of all abutments, 24% of all piers and 13% of all soffits. The most severe contamination was due to leaking deck joints. The spray from traffic under the bridges caused contamination of piers and abutments and, to a lesser extent, soffits.

The study confirmed the value of half-cell potential testing to find areas of likely corrosion, generally where potentials become more negative, but it did not confirm the unlikelihood of corrosion where values were less negative than −200 mV or the near certainty where they were less negative than −350 mV. Chloride contamination was related to reinforcement corrosion but no threshold value was established.

Sulphate content was very seldom greater than 6% by mass of cement. As there was a slight correlation between sulphate and chloride when the latter exceeded 1.5% by mass of cement, it was concluded that some of the sulphate came from deicing salt, although this was not the major source.

A single core was cut from an abutment or pier of 100 of the bridges. Although most bridges contained potentially reactive aggregates, only one third of them showed any sign of alkali-silica reaction. There was a slightly higher incidence in Devon and the West Midlands, as expected.

The 200 bridges were classified on the basis of the extent and severity of visual defects as follows:

- ☐ 25 were good,
- ☐ 114 were fair,
- ☐ 61 were poor, although none was unsafe.

Three-quarters of all deck joints were leaking, thereby tending to increase chloride contamination.

The most common cause of cracking was early thermal effects.

The bridges were also classified on the basis of half-cell potential and chloride concentrations. Many that appeared to be in poor condition visually were fair when tested. Half of those classed poor when tested had been classed good or fair from appearance. Surface defects did exist without other causes of corrosion, such as chloride contamination. However, corrosion of the reinforcement may have started without producing any visible signs on the surface of the concrete.

6.1.10 Bridges in the former Yugoslavia—Radić (1989)

Durability data on bridges in an aggressive environment was gathered in the first comprehensive inspection of 2210 bridges undertaken in the late 1980s in Croatia. Many of the bridges were built between 1945 and 1975, including the following long span concrete arches:

Sibenik Bridge	251 m	1966
Land-Pag Island Bridge	198 m	1968
Titov Bridge	250 m and 390 m	1980

The condition survey resulted in the following percentages.

Good	25
Minor deficiencies	56
Major deficiencies	13
Worn out	6

The percentages of deficiencies due to various causes were as follows.

Design	27
Construction	41
Lack of maintenance	32
Other causes	10

The overlap of the latter figures reflects the interaction of some of the causes of deficiencies.

6.1.11 Concrete bridges in Northern Ireland—Montgomery and Murray (1990)

Of the 7700 road bridges in Northern Ireland, approximately 1300 are of reinforced concrete construction. Some date back to the early years of this century but many large reinforced concrete bridges were constructed in the 1960s and 1970s during the development of the motorway system. In 1987, the Department of the Environment (Northern Ireland) commissioned a survey of the condition of the bridges leading to proposals for a programme of protective measures. This concentrated on surface treatments and cathodic protection.

6.1.12 Bridges in Czechoslovakia—Dahinter (1990)

There are about 46 000 highway bridges in Czechoslovakia with spans of more than 2 m. Approximately two-thirds are plain concrete, reinforced concrete or prestressed concrete. The earliest were built at the beginning of this century, most of the reinforced concrete bridges between the First and Second World Wars, and the prestressed superstructures since 1950. The management of all highway bridges is controlled by two Standards. Experience and results of investigations are described and the complex diagnostic method is explained with examples.

6.1.13 Condition of Indian bridges—Merani (1990)

A condition survey of bridges on National Highways in India carried out in the late 1980s found that, of the 6500 bridges in 23 States, about 20% showed distress, nearly half of these being in need of replacement. A similar survey in Maharashtra State on State Roads found 871 bridges needing repairs and strengthening. This led to new proposals for the inspection, maintenance and rehabilitation of bridges.

6.1.14 Viaducts in Italy after 25 years—Petrangeli (1990)

This paper reports on the inspection and repair of a number of viaducts on the A3 highway linking Central to Southern Italy, with particular reference to dynamic tests which yielded data on the effective prestressing forces acting on the beams. The structures were in bad condition due to the extensive use of deicing salt and the absence of waterproofing. A dramatic loss of prestress was found and remedied by the addition of new prestressing cables. Improved seismic behaviour was obtained by making the deck slab continuous over the supports and by adding energy dissipating devices at the abutments.

6.1.15 British highway bridges—Mays (ed.) and Boam (1992)

Chloride contamination, due to the use of road salt, is considered to be the most serious cause of the corrosion problem affecting at least 60 000 highway bridges in which concrete is the main structural material of:

- deck
- substructure
- retaining walls.

Proper examination and correct diagnosis of the cause (or causes) of deterioration are essential for the correct specification of repairs. Six options can be considered:

- do nothing
- reduce corrosion rate
- repair visible defects
- perform major repairs
- cathodic protection
- replace affected elements.

The approach is influenced by cost and each case must be considered separately.

6.1.16 British railway bridges—Mays (ed.) and Dixon (1992)

Durability problems of the 3800 British concrete railway bridges arise mainly from reinforcement corrosion and concrete deterioration due to chlorides or carbonation. The variety of structures located in all parts of Britain is such that many types of environment occur and all of the known deterioration mechanisms are involved. British Rail Research developed a portable kit for on-site testing of chloride concentrations and carbonation depths. Case studies are given where the dominant cause was:

- chloride
- carbonation
- alkali-silica reaction
- frost damage, or
- poor surface finish.

British Rail's durability research is mentioned.

6.2 Performance of repairs

6.2.1 Effect of repairs on corrosion—Boyd, Lankard and Slater (1979)

Basic concrete specimens 711 × 711 × 203 mm had voids which were filled using various repair techniques. Corrosion of reinforcement adjacent to the 'repair' was monitored for 18 months with periodical applications of deicing salt. Findings supported the use of membranes.

6.2.2 Injected cracks—Calder (1986)

In investigation of structural performance, slabs having cracks from 0.1 to 0.4 mm wide were injected with epoxy resin by a specialist contractor. When recracked, only 30% of the new cracks were associated with repaired cracks. The compressive strength of the repaired prisms in slant shear tests averaged 95% of the uncracked prisms. Similar tests using polyester and methylmethacrylate resins were less satisfactory, which could have been partly due to their poorer penetration. Tests on the slabs treated with a liquid silicate solution showed that they had no structural strength.

6.2.3 Durability of repaired slabs—Calder and Thompson (1988)

Carbonation in unrepaired slabs reached behind the reinforcement compared with less than the cover in the slabs repaired with epoxy, polyester and methylmethacrylate resins. These repairs prevented preferential access of sodium chloride into the cracks but did not prevent penetration of chloride ions from the surface. About 80% of all corrosion sites were at cracks. Repairs reduced the total number of corrosion sites by about half but had little effect on the number of sites with some loss of section. There was little to choose between the resins in improving durability but the silicate was ineffective.

6.2.4 Protection afforded by crack injection—Calder (1990)

Cracked reinforced concrete slabs were repaired with:

- epoxy resin
- polyester resin
- methylmethacrylate resin
- liquid silicate solution.

Comparisons with unrepaired slabs after ponding with salt water for periods up to three years led to the following firmer conclusions.

- About 80% of corrosion was at cracks.
- Ranking penetration in descending order was

  epoxy (best) > polyester > MMA (least good)

- Carbonation extended beyond the reinforcement in the unrepaired slabs. Each of the three resins reduced the carbonation depths to less than the cover. Silicate achieved very little improvement.
- Salt water readily penetrated unrepaired cracks and caused corrosion at about half of them.

6.3 Performance of surface treatments

6.3.1 Berlin bridges—*New Civil Engineer* (1985)

The severe winter conditions in Berlin led to the spread of 60–80 t of salt per kilometre, more than ten times the rate of application in the UK. The Berlin Senate had set up a department specifically to try out and test new waterproofing products to safeguard their 500 concrete bridges. They had checked chloride and carbonation in good quality concrete after 15 years' exposure. Their results are given in Table 6.4.

Table 6.4. Penetration depths after 15 years: in mm

	Chloride	Carbonation
Abutments and piers	100	40–60
Deck soffits	10–20	10

Daily temperature variation could result in crack widths varying as much as 0.4 mm over twelve hours. Research had led to the specification of the following treatments.

Upper surface of deck	sprayed seamless membrane
Superstructure sides and soffit	silane + acrylic coating
Substructure exposed to splash	elastic membrane
Substructure below ground	bituminous coating

The silane impregnation formed a barrier against chloride and the acrylic against carbonation, while allowing moisture to escape from the concrete. It was claimed that silane would last as long as the bridge but this claim was based on only twelve years of observation. Acrylic coatings had to be renewed after ten years.

As silanes need a pH of 9 or more to form siloxane compounds, siloxanes were used for the treatment of carbonated concrete. Preparation of the concrete surface by hot water spray was important, using pressures up to 10 N/mm^2 for older concrete.

6.3.2 Effectiveness of silane for reducing corrosion—Hammersley, Dill and Darby (1990); Darby (1991, 1992)

An investigation was undertaken on Wolvercote Viaduct, Oxford, into the effectiveness of silane for reducing chloride-induced corrosion activity. The trial was in three stages:

- ☐ survey
- ☐ remedial work and instrumentation
- ☐ monitoring.

■ The survey in May 1988 comprised:

- ☐ visual inspection and hammer sounding,
- ☐ cover survey,
- ☐ half-cell potential mapping using silver/silver chloride reference electrode (SSCE),
- ☐ sampling and analysis for chloride concentrations,
- ☐ carbonation depth testing.

The chloride concentrations at the depth of the reinforcement (30–60 mm) commonly exceeded 1% by weight of cement. Half-cell potential mapping showed wide variations with extreme values of −540 mV (SSCE). Carbonation depths were less than 5 mm.

■ Remedial work carried out June-August 1988 consisted of:

- ☐ cutting out areas of cracking and spalling leaving visibly intact chloride-contaminated concrete in place,
- ☐ grit blast cleaning reinforcement,
- ☐ patch repairs with cementitious mortar,
- ☐ making tops of beams free-draining,
- ☐ steam cleaning concrete surfaces,
- ☐ application of silane to sides and soffits of beams,
- ☐ application of acrylic waterproofing to tops of beams,
- ☐ completion of drainage outlets.

■ Nine months later, half-cell potential mapping showed some reduction in potentials and signs of new 'high spots' adjacent to repair areas. Eighteen months after the completion of the remedial work, the reduction in potentials was generally between 50 and 150 mV, except for one large repair area where the potential had reduced from −560 mV to −150 mV. The authors concluded (1990) that the concrete had started to dry out.

Darby (1991) gave a State-of-the-Art Review of impregnation treatment to counter chloride attack, noting the scale of the operation and some problems in its implementation. The County Surveyor's Society had found in 1974 that 72% of counties had confirmed cases of significant chloride attack. Despite some difficulties with the instrumentation at Wolvercote Viaduct, Darby was able to confirm that the resistivity of the concrete had increased and that drying out was continuing. The success of impregnation was very dependent on the surface condition at the time, as affected by the weather and the cleaning method, and rates of application which were difficult to supervise.

Darby (1992) reported an extension of the trial to three more beams. Resistivity had increased by 62–233%. Potentials had become less negative by 28–122%.

6.3.3 Further studies of impregnation—Brown (1990) and Pearson-Kirk (1992)

The studies by Satoh *et al.* (1989) are summarized in 5.2.4.

Brown (1990) described a number of trials commissioned by the Scottish Development Department to establish the effectiveness of silane. The capping beams of Cromarty Bridge had deficient cover and substantial chloride ingress; chloride concentration was high in the splash zone of columns. Half-cell potentials suggested a high probability of corrosion. An area was treated with two applications of silane and cores were taken after a few weeks, 6 months and 33 months to compare with measurements of half-cell potentials, chloride concentrations and depths of carbonation. The penetration of the silane ranged from less than 0.5 to 2 mm; it reduced water absorption by 90% but the capillary porosity was only reduced by 25%. Although chloride concentrations remained the same in the treated and untreated areas, half-cell potentials were found to be decreasing in the treated areas.

Pearson-Kirk (1992) reported some results from monitoring four structures on the Charmouth Bypass and eight on the A303 Mere to Wincanton improvement. Shortly after construction, water absorption was measured before impregnation with silane, then between applications and again after the second application. Depths of penetration did not exceed 1 mm. Water absorption was reduced by 60–75% by the first application and a further 7–15% by the second.

6.4 Cathodic protection

6.4.1 Methods of corrosion control—McKenzie (1986)

The effectiveness of two methods was reported.

- *Enclosure*. The steel beams of three composite highway bridges were enclosed to provide a protected environment. Monitoring over periods up to five years showed that the enclosed environment inhibited corrosion of bare steel and kept painted steel maintenance-free for decades. Soffits of the composite deck slabs are also protected.

- *Cathodic protection.* Tests were carried out on concrete units containing reinforcement manufactured as a cubic lattice of blast cleaned mild steel. Half of the units contained added chloride to accelerate corrosion. Levels of protection ranged from zero to over-protection, using impressed currents through noble metal feed wires to conductive coatings on one vertical and one horizontal surface. The specimens were exposed to the atmosphere and salt was applied each week. Early results suggested that currents had spread to all parts of the reinforcement at levels to provide protection.

6.4.2 Cathodic protection in Missouri—Girard (1987)

Cathodic protection had developed rapidly and had been applied to box beam, box girder, voided and flat slab bridges, all in need of roadway surface repair. Various systems had been used in these rehabilitation attempts and experience showed the need for close inspection during installation and continued monitoring and evaluation during operation. Maintenance may not totally cease simply as a result of cathodic protection.

6.4.3 Assessing cathodic protection—Lewis and Boam (1987)

In view of the scale of deterioration of bridges in the UK due to chlorides, work was in progress to assess the effectiveness and cost of cathodic protection as a means of extending the life of affected structures. At a stage where only superficial repair appeared to be required but where significant chloride-induced corrosion existed and was progressive, some motorway viaducts were being used in field trials. Reference was made to the research noted in 6.4.1.

6.4.4 Trial installation on viaduct—Romagnolo, Ferrari and Grandi (1988)

A description was given of an advanced cathodic protection system with computer controlled monitoring, photoelectric feed and special coating grouts. This first application in Italy, on an Apennine motorway viaduct between Florence and Bologna, provided the opportunity to study:

- ☐ dispersion of salt from the concrete
- ☐ protection against stray currents
- ☐ need for deck waterproofing, and
- ☐ the possibility of long-term monitoring.

6.4.5 Applications of CP in Scandinavia and the UK—Berthagen (1990) and Wyatt (1990)

Berthagen (1990) reported that the first test installations in Denmark, Sweden and Norway were completed in 1987 and 1988 and in Finland in 1989. During

1988 a cathodic protection system was applied to four columns of the Öland Bridge (see 2.3.11) which had been repaired by different methods. A system was installed on the underwater part of the Smögen Bridge on the west coast of Sweden.

Wyatt (1990), reviewing applications in the UK, referred to earlier descriptions of cathodic protection systems (1987 and 1989) and quoted from Wallbank (1989):

> 'The estimates assume that cathodic protection would be used widely as a repair technique. Although further research and trials are required, the potential economic benefits of cathodic protection are so great that it is recommended that the techniques be adopted on a wide variety of bridges. Such use would aid the rapid collection of data concerning its application and durability. Furthermore, even if protection were found only to delay the need for major repairs or replacement by a few years, it would probably be justified economically. It is estimated that the use of cathodic protection on 14 of the 200 bridges in this study would cost about £4,650,000 less than the alternative of cutting out and repairing the concrete; over £300,000 per bridge'.

The first trials by the Department of Transport were on the Midland Links (see 2.3.6) at Gravelly Hill in 1986. Wyatt reported that the first fully commercial system in the UK was installed in 1987 on a road ferry access ramp and, in the same year, the first UK bridge deck was cathodically protected.

6.4.6 A water-based conductive coating for piers—Clemena and Jackson (1991)

A water-based conductive coating was applied as the anode in two cathodic protection installations together with a remote monitoring system using a microprocessor.

6.5 Special techniques

6.5.1 External bonded reinforcement—Calder (1989)

Exposure tests on externally reinforced concrete beams were noted in 4.4.

6.5.2 Bridge deck waterproofing—Price (1989, 1991)

The trials (1989) covered a wide range of 25 sheet, roll and board waterproofing systems and 22 liquid systems applied to the concrete surface. Manufacturers installed their own systems onto their allotted bays of the concrete test slab. Their performance was observed during installation and curing. The systems were left exposed for about four weeks and were then overlaid with asphalt surfacing. Performance of the systems was monitored for three years and then they were inspected after removal of the surfacing materials. Many of the systems were damaged during the laying of base course and all the specific shortcomings were presented for the different systems, classified according to their principal base materials.

Price (1991) reported on current practices in installing, repairing and renewing bridge deck waterproofing systems and related these practices to in-service failures.

6.5.3 Fibre shotcrete in the UK—Robins (1990)

Possible advantages of using fibres were given as:

- post-crack resistance or toughness
- higher impact resistance
- improved flexural strength
- saving in cost of fixing mesh reinforcement
- faster construction
- reduced thickness.

Possible disadvantages were:

- high cost of fibres
- discomfort of operatives
- occasional difficulty in mixing and spraying
- concerns (often unjustified) about durability.

The usual reason for using fibres was to improve energy absorption, although there were some gains in strength and possibly in construction. The vast majority of applications in the UK used steel fibres, either melt extract or hooked end wire, while enlarged end slit sheet fibres were popular in Scandinavia and Australia. Descriptions were given of steel fibre applications by British Rail in a tunnel and on three bridges; also guidance on specifications (see 7.4.4).

6.5.4 Seven lightweight aggregate concrete bridges—Mays and Barnes (1991)

Seven case histories are given of bridges constructed between 1969 and 1977 (Appendix 2) with results of recent inspections. Redesdale Bridge, built in 1969, has been the subject of continuous monitoring by TRRL as part of a study of creep and shrinkage. Despite some slight honeycombing at the bottom of webs of prestressed beams there is no sign of deterioration. There is no evidence that lightweight aggregate concrete is any less durable than normal-weight concrete. None of the bridges was in need of any repair.

6.5.5 Flowable concrete—Patel (1992)

Three proprietary flowable concretes were assessed using the tests specified by the Department of Transport (1986). Aspects investigated were:

- effects on the flow properties of temperature and water content,
- variation of the flow properties between batches,
- effects of water/material ratio on shrinkage and pull-off strength.

The mixing temperature had a significant effect on the flow test results. The temperature at which the concrete was poured had little effect on the flow, provided the materials were stored and mixed at or above 15°C. Flow properties improved with increasing water content up to a critical level. The wide variations between batches confirmed the need to test every batch before use.

6.6 References

Berthagen L. (1990). Cathodic protection of steel in concrete structures in Scandinavia. *British–Swedish Seminar on Concrete Repairs*. Chalmers University of Technology, Goteborg.

Bickley J. A., Emery J. J., Sturrup V. and MacDonald P. (1986). *Repair techniques for concrete bridge components*. Roads and Transportation Association of Canada. ISBN 0 919 098 48 7.

Booth, Sweetman and Wolfe (1987). *Bridge maintenance survey results*. Research Project 108502 on behalf of the Bridge Maintenance Coordinating Committee, National Roads Board, New Zealand.

Boyd W. K., Lankard D. R. and Slater J. E. (1979). *Influence of bridge deck repairs on corrosion of reinforcing steel*. Final Report. Transportation Research Board, Washington DC, USA.

Brown J. H. (1986). *The performance of concrete in practice*. A field study of highway bridges. TRRL Contractor Report 43.

Brown B. J. and Cochrane S. (1989). Condition survey of seven bridges, M8 motorway, Renfrew Bypass. *Proc. Int. Conf. on Structural Faults and Repair*. Vol. 2, Kensington, London, 27–29 June.

Brown B. J. (1990). R. Dhir and J.W. Green (eds) Silane treatment of concrete bridges. *Proc. Int. Conf. on Protection of Concrete*. University of Dundee, 11–13 Sept., 1059–1080.

Calder A. J. J. (1986). *Repair of cracked reinforced concrete: assessment of injection methods*. TRRL Research Report 81.

Calder A. J. J. (1988). *Exposure tests on externally reinforced concrete beams—performance after 10 years*. TRRL Research Report 129.

Calder A. J. J. and Thompson D. M. (1988). *Repair of cracked reinforced concrete: assessment of corrosion protection*. TRRL Research Report 150.

Calder A. J. J. (1989). *Exposure tests on 3.5 m externally reinforced concrete beams—the first 8 years*. TRRL Research Report 191.

Calder A. J. J. (1990). Corrosion protection afforded by various crack injection techniques. *Proc. 3rd Int. Conf. on Corrosion of Reinforcement in Concrete Construction*. Wishaw, Warwickshire, 21–24 May. Published for the Society of Chemical Industry by Elsevier Applied Science, 433–445.

Clemena G. G. and Jackson D. R. (1991). Cathodic protection of the concrete piers of two bridges in Virginia using a water-based conductive coating. Durability of concrete. *Proc. 2nd Int. Conf., Montreal*. ACI Special Publication SP-126, 679–696.

Dahinter K. (1990). Inspection and strength evaluation of concrete highway bridges in Czechoslovakia. Illustrative examples. *Proc. 1st Int. Conf. on Bridge Management*. University of Surrey, Guildford, 28–30 Mar., 386–393.

Darby J. J. (1991). *Impregnation of concrete structures*. Department of Transport Maintenance Engineers Conference.

Darby J. J. (1992). Maintenance—a County viewpoint. *Durability of Highway Structures Symposium*. Institution of Highways and Transportation, Taunton, 12 Nov.

ELLIS W. J. and BIANCHETTI R. L. (1979). *Corrosion control and repair of concrete bridge structures.* State-of-the-Art Report.

GIRARD R. J. (1981). *Experience with cathodic protection in Missouri.* Transportation Research Record No. 1113, 39–46.

GOEHLER B. (1990). Experiences with the first generation of prestressed concrete bridges in Germany. *Proc. 1st Int. Conf. on Bridge Management.* University of Surrey, Guildford, 28–30 Mar., 287–294.

GORDON S. (1989). Durability of highway bridges. *Symp. on the Durability of Structures.* Lisbon. International Association for Bridge and Structural Engineering.

HAMMERSLEY G. P., DILL M. J. and DARBY J. J. (1990). An investigation into the effectiveness of silane for reducing corrosion activity in a chloride-contaminated reinforced concrete bridge structure. *Proc. 1st Int. Conf. on Bridge Management.* University of Surrey, Guildford, 28–30 Mar., 655–666.

HARM E. E. (1993). Private communication. Illinois Department of Transportation.

HAY R. E. and VIRMANI Y. P. (1985). North American experience in concrete bridge deterioration and maintenance. *Proc. of a one-day Symposium.* London Tara Hotel, Concrete Society, 25 Sept. ISBN 0 946691 09 6.

INGVARSSON H. and WESTERBERG B. (1985). *State-of-the-Art Report concerning durability, maintenance and repair of bridges and similar structures.* Vaeg och Trafik Institut Medd 449, 115.

INGVARSSON H. (1988). Operation and maintenance of bridges. *Pre-Conf. Proc. 3rd IRF Middle East Regional Meeting: Towards Better Road Performance.* Riyadh. 13–18 Jan., **6**, 6.31–6.58. International Road Federation, Washington DC, USA.

INGVARSSON H. (1990). Maintenance and repair need of Swedish bridges. *British-Swedish Seminar on Concrete Repairs.* Chalmers University of Technology, Goteborg.

LACROIX J. E. (1985). *Bridge deck condition survey.* Phase V. Long-term performance evaluation. Final Report. Illinois Department of Transportation, USA.

LEBEK D. (1985). German experience in concrete bridge deterioration and maintenance. *Proc. of a one-day Symposium Concrete Bridges—Investigation, Maintenance and Repair.* London Tara Hotel. Concrete Society, 25 Sept. ISBN 0 946691 09 6.

LEWIS D. A. and BOAM K. J. (1987). Cathodic protection of reinforced concrete. *Proc. Int. Conf. on Deterioration and Repair of Reinforced Concrete in the Arabian Gulf.* Bahrain Society of Engineers, Vol. 1, 79–97.

MCKENZIE M. (1986). *Alternative methods of bridge corrosion control—enclosure and cathodic protection.* Institution of Highways and Transportation National Workshop, Leamington Spa.

MAYS G. C. and BARNES R. A. (1991). The performance of lightweight aggregate concrete structures in service. *The Structural Engineer,* 15 Oct., **69**, No. 20, 351–361.

MAYS G. C. and BOAM K. J. (1992). Highway bridges. *Durability of concrete structures. Investigation, repair, protection.* E. and F. N. Spon, 188–205.

MAYS G. C. and DIXON J. F. (1992). Railway bridges. *Durability of concrete structures. Investigation, repair, protection.* E. and F. N. Spon, 206–225.

MERANI N. V. (1990). Investigation and rehabilitation of distressed concrete bridges. *Indian Roads Congress Journal,* Nov., **51**, No. 3, 843–871.

MONTGOMERY F. R. and MURRAY A. M. C. C. (1990). Reinforced concrete
bridge protection in Northern Ireland. *Proc. 1st Int. Conf. on Bridge Management.* University of Surrey, Guildford, 28–30 Mar., 259–265.

NEW CIVIL ENGINEER (1983). Maintenance. Swiss work to beat bridge decay. *New Civil Engineer,* 17 Nov. 1983, 26–30.

NEW CIVIL ENGINEER (1985). Berlin bridges take cover. *New Civil Engineer,* 9 May 1985.

PATEL R. G. (1992). *Research on flowable concrete.* TRL Research Report 322.

PEARSON-KIRK D. (1992). Impregnation of concrete. *Durability of Highway Structures Symposium*. Institution of Highways and Transportation, Taunton, 12 Nov.
PETRANGELI M. P. (1990). Inspection and repair of some highway bridges in Italy. *Proc. 1st Int. Conf. on Bridge Management*. University of Surrey, Guildford, 28–30 Mar., 373–381.
PRICE A. R. (1989). *A field trial of waterproofing systems for concrete bridge decks*. TRRL Research Report 185.
PRICE A. R. (1991). *Waterproofing of concrete bridge decks: site practice and failures*. TRRL Research Report 317.
RADIĆ J. (1989). Bridge durability parameters. *Symp. on the Durability of Structures*. Lisbon. International Association for Bridge and Structural Engineering.
ROBINS P. (1990). Fibre shotcrete repair applications in the UK. *British-Swedish Seminar on Concrete Repairs*. Chalmers University of Technology, Goteborg.
ROMAGNOLO M., FERRARI S. and GRANDI M. (1988). Cathodic protection of reinforced concrete structures. Experimental installation on a viaduct in service (in Italian). *Autostrade*, 10/11/12, **30**, No. 4, 77–90.
ROWE G. H., FREITAG S. A. and STEWART P. F. (1984). *Concrete quality in bridges: New Zealand*. Central Laboratories Report 4–84/4.
VAN BEGIN C. (1987). Durability of concrete bridges in Belgium. Balance of the systematic inspection. Atlanta, Georgia. *Katharine and Bryant Mather Int. Conf.*, 27 Apr.-1 May. ACI Special Publication SP100, **1**, 541–554.
VASSIE P. (1986). *Corrosion of reinforcement: an assessment of twelve concrete bridges after 50 years service*. TRRL Research Report 78.
VASSIE P. (1987). *The chloride concentration and resistivity of eight reinforced concrete bridge decks after 50 years service*. TRRL Research Report 93.
WALLBANK E. J. (1989). Performance of concrete in bridges—a study of 200 trunk road bridges. HMSO, London. ISBN 011 550877 5.
WYATT B. S. and IRVINE D. J. (1987). *Materials Performance*, **26**, No. 12, 1221.
WYATT B. S. (1989). Cathodic protection theory and practice. *Proc. 2nd Int. Conf. on Cathodic Protection*, Stratford upon Avon, June.
WYATT B. S. (1990). Cathodic protection of highway structures in UK. *British-Swedish Seminar on Concrete Repairs*. Chalmers University of Technology, Goteborg.

7

Performance and specifications

7.1 Performance

7.1.1 Durability

Beeby (1990) noted the massive growth in the number of papers published annually on the durability of concrete; it doubled between 1976 and 1982 and had doubled again by 1986. Some of the examples noted in preceding chapters lead to the following conclusions.

In North America and Northern Europe, the dominant cause of the deterioration of concrete bridges was the corrosion of reinforcement due to chlorides. More generally, the commonest visible signs of deterioration were cracking and spalling. Shortcomings could usually be attributed to design details, to site practices or to an aggressive environment or combinations of these.

Irwin and Chamberlin (1981) reported on the performance after four and five years of fifteen bridge decks built with 75 mm cover. There was no spalling or delamination and half-cell potentials more negative than -350 mV (CSE) were found in only two spans.

Brown (1986) found the performance of modern bridges to be superior to that of old bridges due generally to better quality and depth of the concrete cover. Although many of the old bridges had areas of low cover and carbonation-related corrosion, many had survived surprisingly well. He considered that carbonation depths in modern bridges were unlikely to exceed 28 mm in 120 years. He was able to relate durability to water/cement ratio or cement content.

Wallbank (1989) reported that some 30% of English motorway and trunk road bridges appeared to be in poor condition on the basis of visual inspection. Classifications on the basis of half-cell potentials and chloride concentration were different in some cases. Three-quarters of all deck joints were leaking. The heaviest chloride concentrations were mostly due to this or to marine exposure. Minimum cover ranged from 0 to 130 mm, with 2% less than 20 mm and 5% greater than 80 mm but was not related to age. Carbonation depths were very variable in bridges of all ages; 90% were less than 5 mm and very few exceeded 20 mm. 72% of the bridges were found to have chloride concentrations exceeding 0.2% by weight of cement.

Vassie (1987) estimated the probability of corrosion to be 10% and 25% where the chloride ion concentrations were 0.2 and 0.4% respectively by weight of cement.

Schiessl (1988) edited the RILEM State-of-the-Art Report on corrosion of reinforcing and prestressing steels in concrete. The Report concentrated on normal corrosion and did not cover hydrogen embrittlement or stress corrosion. The processes leading to depassivation and mechanisms decisive for the corrosion process were described. Advice was included on the durability aspects of concrete mix design, use of corrosion inhibitors, coatings, corrosion-resistant alloys, cathodic protection. Other causes of deterioration were freeze/thaw, alkali-silica reaction, sulphate attack and overloading. An overview of corrosion monitoring techniques was presented.

Macmillan and Treadaway (1990) described a ten-year exposure trial to determine the factors which control durability in simulated Arabian Gulf conditions. Figure 7.1 illustrates the performance of the concrete after the first five years.

Figure 7.1 shows that corrosion initiation is increasingly dominated by carbonation where chloride concentrations are low. With zero chloride, none of the specimens corroded until carbonation reached the reinforcement. When the chloride concentration was 3.18%, corrosion occurred in every case before carbonation reached the full depth of cover.

Lightweight aggregate concrete (see 6.5.4) is no less durable than normal weight concrete.

Thomas and Matthews (1991) reported the findings of an investigation of

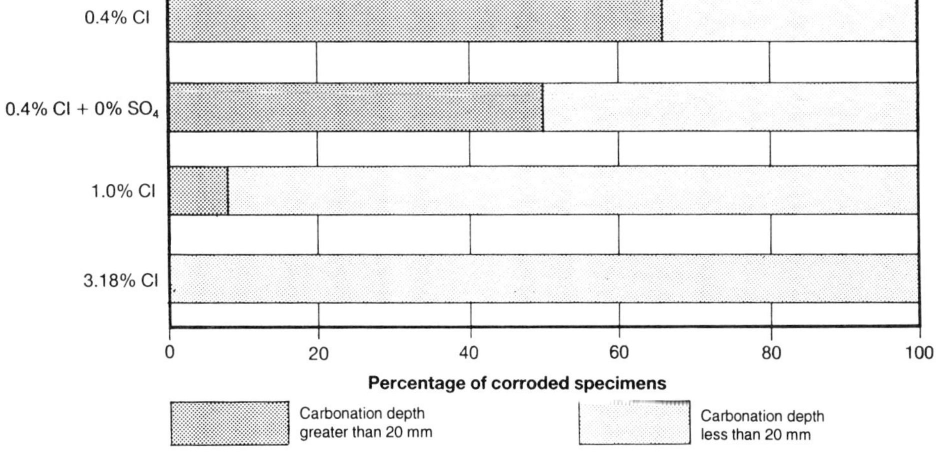

Fig. 7.1. Effects of contaminants on corrosion (Macmillan and Treadaway, 1990)

structures built using PFA concrete. The age of the structures ranged from 2 to 33 years and, at each site, comparable OPC concrete was sampled. Generally, PFA concretes had greater strength, lower permeability, greater resistance to the ingress of chlorides and better suppression of alkali-silica reaction.

7.1.2 Detection of corrosion

Corrosion is usually detected in the following sequence:

- general inspection (visual)
- principal inspection (visual)
- special inspection (sampling and testing)
- monitoring (visual and/or special equipment).

Lambert and Page (1986) described three electrochemical monitoring techniques to make rapid assessments of the condition of reinforcement:

- half-cell potential mapping
- AC and DC polarization of the steel
- electrochemical noise.

Arya and Newman (1990) discussed the three current approaches to establish chloride content, namely:

- total chloride content
- water-soluble chloride content
- free chloride/hydroxyl ion ratio.

Pointing out shortcomings in these methods of predicting the risk of corrosion, they proposed the pore press derivative method which determines the free chloride content.

7.1.3 Performance of repairs

Emberson and Mays (1990) reported on the properties of the various patch repair materials listed in 4.1. Within some of the generic groups there are numerous formulations and fairly wide variations in physical and mechanical properties. Because the samples were chosen to obtain typical characteristics, the results are not to be taken as limiting values. The following summary shows broadly how these material properties compared with those of typical bridge concrete. All of the materials gave adequate compressive and tensile strengths for typical applications.

- Resin mortars had high flexural and tensile strengths but tended to have lower elastic moduli. The resins had much higher coefficients of thermal expansion/contraction than the cementitious materials. They all had greater tensile adhesive strengths than the substrate concrete. Epoxy (A) and acrylic (C) resins shrank slightly within the first few hours of placing and had by far the lowest long-term shrinkage. The temperature of polyester (B) resin rose rapidly and the whole of its high shrinkage occurred within the first 15–20 minutes after placing. Acrylic (C) resin had by far the highest compression creep strain.

- Polymer-modified cementitious materials had coefficients of thermal expansion/contraction much the same as those of many unmodified concretes. The magnesium phosphate material (F) showed a rise in temperature in the first three-quarters of an hour and expanded considerably over the first hour and a quarter. This expansion continued for a month and then the material started to shrink. Tension adhesion failures were at interfaces, including those of SBR (D) and vinyl acetate (E) with their associated primers. Compression creep strain for vinyl acetate (E) was second only to the acrylic resin (C). SBR-modified mortar (D) had an exceptionally low net creep strain which commends its use for repairs to older bridges provided its early shrinkage can be accommodated.
- Cementitious materials started with the obvious advantage of better compatibility with the substrate concrete. OPC/sand mortar (G), the traditional repair material, had the lowest coefficient of thermal expansion/contraction and one of the lowest values of net creep. Its tension adhesion failures were predominantly in the concrete substrate and this was entirely so for flowing concrete. The flowing concrete (I) equalled the epoxy resin (A) with the highest average pull-off load.

All materials were judged to be suitable for repairs on the basis of strength but this was qualified by the potential load redistribution in those with a mismatch of elastic moduli.

Beeby (1978 and 1983) and Manning (1985) recognized that cracks in concrete, regardless of their width at the surface, provide preferential means of ingress to the reinforcement for aggressive agents. The Concrete Society (1982 and 1984) and the Department of Transport (1986a and 1990a) pointed out the need to establish the cause of cracking before attempting a repair. After the cause has been remedied or monitoring has confirmed that movement has ceased, cracks wider than 0.3 mm may be injected and areas with finer cracking may be protected by impregnation to improve durability.

Calder (1986) reported the structural performance of cracked reinforced concrete specimens repaired with four materials. The results of the tests, noted in 6.2.2 and set out more fully in Table 7.1, suggest that the epoxy resin was the most effective structural repair and that the silicate was ineffective.

Table 7.1. Structural performance of crack repairs

Material	Slant shear test strength / Uncracked strength	New cracks at repairs
Epoxy resin	95	30
Liquid silicate	—	90
Polyester	60	60
Methylmethacrylate	75	90

The stiffness of a cracked slab was about one-quarter of that of an uncracked slab. All of the repairs restored about half of this loss.

Calder and Thompson (1988) described the second stage of the above research, noted in 6.2.3, which was to assess the corrosion protection afforded by the different crack repairs. Results after exposures up to 72 weeks were noted in 6.6.3.

Calder (1990) gave results after three years' exposure. (See 6.6.4.)

Cairns (1990) presented a non-linear analytical model to study redistribution of stress in a reinforced concrete beam after repair and the effects of relieving dead loads during repair.

Keer, Chadwick and Thompson (1990) reported on the protection afforded to reinforcement by twelve repair materials compared with control specimens of good and moderate quality concrete. They determined the following ranking in increasing time to corrosion at 10 mm cover, poorest first:

(*a*) moderate concrete
(*b*) silicate-modified mortar
(*c*) methyl methacrylate mortar
(*d*) 10 mm aggregate concrete
(*e*) good concrete
(*f*) PFA concrete
(*g*) sand/cement mortar
(*h*) GGBFS concrete.

The following showed no corrosion after 79 days or more:

(*i*) flowing concrete
(*j*) microsilica concrete
(*k*) acrylic-modified mortar
(*l*) SBR-modified mortar
(*m*) epoxy mortar
(*n*) lightweight epoxy mortar.

At Haddiscoe Bridge (2.2.4), the mastic asphalt waterproofing restricted transport of deicing salt into the deck concrete considerably, but not totally. Repairs of 75 mm depth performed well with no new corrosion. Repairs of 25 mm depth tended to delaminate around their perimeter. (For criteria see 7.3.4.)

7.1.4 Performance of surface treatments

As noted in 5.2.1, Pfeiffer and Scali, in NCHRP Report 244, (1981) found the following five surface treatments superior to others on the market:

- epoxy with 50% solids
- epoxy with 100% solids
- methyl methacrylate
- moist-cured urethane
- alkyl-alkoxy silane.

In soaking and drying tests, these reduced chloride ion content by 79–97% compared with uncoated concrete. Epoxies generally had the better adhesion and were expected to last about ten years.

As noted in 5.2.4 and 6.3.1/2/3, alkyl-alkoxy silane has been used widely in Europe. Successful impregnation with silanes depends upon the concrete surface being clean and dry. Penetration depths of 3 to 5 mm were reported by Satoh (1989), achieving reductions in water absorption of 75–80% and similar results were reported by Brown (1990) and Pearson-Kirk (1992) where penetration depths were less than 2 mm. Darby (1992) reported results of monitoring half-cell potentials and resistivity in a silane-treated concrete viaduct. In the first eighteen months the treatment was proved to be effective in drying out the concrete and, after four years, resistivities had increased and negative potentials had decreased very significantly.

In 5.2.5 it was noted that silane effectively repelled water, thus aiding the drying out of concrete affected by alkali-silica reaction. Miyagawa *et al.* (1991), Blight (1991) and Kojima *et al.* (1992) had studied this effect.

Harwood (1989) discussed criteria for specification of surface coatings, especially for the Arabian Gulf environment where relative humidity varies up to 79% and temperatures range from 20–35°C.

Thompson and Leeming (1992) reported water vapour, water and chloride ion transmission of 21 surface treatments and postulated the following general performance requirements:

(*a*) applicability and tolerance to temperature and moisture
(*b*) good adhesion or penetration
(*c*) acceptable appearance
(*d*) long-term stability and tolerance to UV radiation
(*e*) long-term resistance to ingress of chloride solution
(*f*) long-term resistance to ingress of water
(*g*) long-term resistance to ingress of carbon dioxide
(*h*) long-term water vapour transmission
(*i*) long-term tolerance to microcracking in substrate
(*j*) ease of maintenance and overcoating.

The study dealt principally with requirements (*e*), chloride resistance, and (*h*), water vapour transmission. The following conclusions were drawn.

(i) All 21 treatments reduced the rate of chloride ion uptake by at least 90%.
(ii) The rate of absorbing chloride solution into concrete was linear with the square root of time, a capillary action.
(iii) Treated specimens dried at rates not exceeding 15% of the control values.
(iv) Artificial weathering tended to reduce chloride uptake and increase drying rates.
(v) Film-forming treatments reduced the rate of water vapour transmission by more than 50%.
(vi) The hydrophobic treatments transmitted water vapour at rates from 50–110% of control values.

(vii) There was considerable variability in the behaviour of specimens, and rankings should be treated with caution.

See 5.3 for performance of corrosion-protected and corrosion-resistant reinforcement in concrete.

7.1.5 Performance of cathodic protection

Cathodic protection was first applied to bridges in the USA in 1968 and the installation at Sly Park Bridge, California, was reported by Stratfull (1974). As noted in 5.4.1 and 5.4.2, efforts in the USA, Canada and elsewhere were concentrated on the development of effective anode systems and monitoring techniques. Criteria for cathodic protection were described by Stratfull (1983) who came to the following conclusions.

(*a*) A potential shift of 400 mV should control 97½% of bridge deck corrosion but the potential shift should not exceed that to cause a most negative potential of -1100 mV CSE.
(*b*) Locations for half-cell potential monitoring should be near locations of the most negative half-cell potential found on the deck.
(*c*) The E log I constant current criterion appears to be an appropriate means of controlling the corrosion of steel in concrete bridge decks.
(*d*) Because half-cell potentials of cathodically protected reinforcement in bridge decks are erratic, requiring all half-cell potentials to be at or above a critical value does not appear to be a suitable method for control.
(*e*) Analysis of data from corrosion test probes showed that the bottom mat of reinforcement received 13% of the current received by the top mat of reinforcement.

From this he inferred that, at low current densities, the steel in the low salt-content concrete, which was cathodic in the corrosion cell, would receive current from the cathodic protection and the steel in the salty concrete, which was anodic in the corrosion cell, would only begin to receive current when its corrosion potential had been neutralized. Thereafter the top mat would receive a far greater current density from the system than the bottom mat.

The considerable experience, in North America, of cathodic protection systems for unwaterproofed and unsurfaced concrete bridge decks, led to its general acceptance there as a fully proven technique.

Cathodic protection has been applied there to substructures since the mid 1980s using conductive paint, sprayed metal coating and conductive mesh anodes. Problems have arisen with poor durability and debonding of overlays.

Reel and Gulis (1993) described applications of cathodic protection to bridge decks and piers in Ontario.

In the UK, waterproofing has generally provided good protection to bridge decks and cathodic protection is not appropriate (see 5.4.1). The major interest lies in its potential applications to substructures. It seems likely that comparatively high installation costs could be outweighed by considerable savings in the extent of physical repairs to chloride-contaminated reinforced concrete.

Research (see 6.4.1) at TRL is already indicating that cathodic protection reduces and may eliminate corrosion in such cases.

The Department of Transport has also set up an investigation into the effectiveness and relative costs of alternative cathodic protection anode systems such as:

- conductive paints
- conductive polymer in an overlay
- conductive tiles with connecting titanium strips
- sprayed zinc coating
- wire mesh with overlay as shown in Figure 5.5.

The PIARC Technical Committee on Bridges (1991) reported cathodic protection installations, on a large scale in Italy and to lesser extents in seven other European countries and Japan. In Italy, most of the installations are on prestressed concrete decks with the possible problem of hydrogen embrittlement of the tendons. Elsewhere the main problem is assurance of effective anodes. The titanium net type has been widely used, but conductive polymer and conductive coating types have also been tried.

7.2 Philosophy

7.2.1 Reliability

Different teams of people are usually responsible for design, construction and maintenance of bridges and each has an effect on its whole life performance, whether good or bad. Whatever feedback and interaction there has been between the different teams hitherto has had little formal structure. Since CIRIA Report 63 (1977), probabilistic methods were adopted for the design of bridges, albeit adopting arbitrary values for the partial factors until the necessary statistical data could be gathered to refine them. The use of reliability analysis for the assessment of durability by Siemes and Vrouwenvelder (1985), the OECD (1989 and 1992) and Miyagawa (1991) has been noted (see 1.1.2).

A number of papers on the subject were presented at the IABSE Symposium on Durability (1989).

Borges (1989), in his opening address, referred to the CEB (1983) Bulletin No. 152 on the durability of concrete structures and the fact that deterioration models had not been established for the important durability mechanisms.

Cabrera, Cusens and Lynsdale (1989) noted that compressive strength, the usual control parameter for concrete quality, is related directly to permeability for a particular concrete but this relationship cannot be generalized to apply to concretes of different composition. Statistical analysis of their data had led to the conclusion that oxygen permeability, K (m^2), is directly related to porosity, P%, and inversely to strength f_{cu} (N/mm^2). They gave mathematical models for both OPC and OPC/PFA concretes, reproduced here as Figure 7.2.

They went on to consider permeability as a control test which would require measurement at early ages during mix design. They gave the following model to

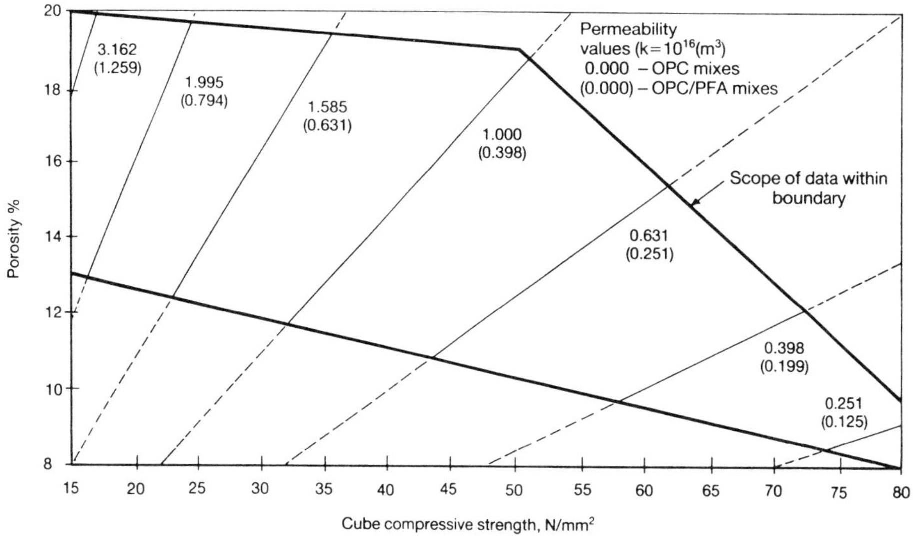

Fig. 7.2. Porosity – strength – permeability model (Cabrera, Cusens and Lynsdale, IABSE, 1989)

predict the permeability of concrete up to 90 days from its one-day value, K_1 (m^2)

$$\log K = -3.688 + 0.789 \log K_1 + 0.356t$$

This led to a criterion for good concrete to have a target value for oxygen permeability at one day of 1×10^{-16} m^2.

Singh (1989) described a stochastic life-expectancy and financial analysis on microcomputer using a Monte Carlo simulation, illustrated with an example of a footbridge.

Andrade, Alonso, Gonzalez and Rodriguez (1989) presented a methodology to relate loss in load-carrying capacity as a function of the rate of corrosion of the reinforcement. However, the assumptions represented carbonation and uniform loss around the whole surface of the steel with no pitting.

Poineau (1989) discussed the assessment of durability from design to execution.

Kunishima and Okamura (1989) introduced the concept of a durability index.

Hergenröder, Schall and Rackwitz (1989) presented stochastic degradation models and discussed the updating of reliability characteristics with data from inspections. The proposals were illustrated by an example of a concrete structure having carbonation-induced spalling.

Wood, Johnson and Ellinas (1990) discussed the application of reliability analysis to evaluate the safety of deteriorating reinforced concrete structures. While CIRIA Report 63 and BS5400 provide a ready made framework with a simplified partial factor approach, no allowance for deterioration was made so it is necessary to look for relevant statistical data.

Fortunately, full scale structural failures are too rare to provide statistical data but the collapse of the Ynys-Y-Gwas Bridge provided a useful source of information. The corrosion of reinforcement or the deterioration of concrete can produce a rapid, catastrophic failure mode comparable to buckling in steel structures. Inspection and/or proof loading give no warning of impending failures of this type. Similarly, alkali-silica reaction could be catastrophic in a critical location especially if this were to coincide with a local leakage of surface water. Punching shear failure is brittle and half-joints provide the classic situation for pitting corrosion.

It is necessary to extend the body of test and case study data in order to apply reliability analysis.

7.2.2 Information systems

It has been noted (1.1.2) that Bridge Management Systems were being developed in the OECD countries and that International Conferences (1.7.2) on the subject were held in 1990 and 1993.

Dawe (1990) described the National Structures Database which holds details of all Department of Transport structures, their location, construction and ancillary components, together with information from inspection reports and details of maintenance expenditure. The costs of remedial work arising from General and Principal Inspections are estimated and the work is given one of five priority ratings as a basis for developing annual regional and national programmes. Largely, but not entirely, due to the increased deterioration of bridges as a result of deicing salt, the Department of Transport has developed a 15-year strategy to strengthen its bridges in accordance with its latest code and to make up other shortcomings where the modern standards are not met. When this programme has been completed, the annual maintenance programmes will cope with the ongoing deterioration.

A Second International Conference on Bridge Management (1993) included papers dealing with the following.

- Design for minimum maintenance
- Protection
- Performance
- Maintenance
- Bridge strengthening
- Non-destructive testing
- Whole life costing
- Service life predictions
- Inspection
- Assessment and evaluation
- Repair and rehabilitation
- Demolition.

The essential link between these aspects of Bridge Management is the need for the effective feedback of data with the means of its storage and retrieval in information systems.

7.2.3 Design life and service life

Somerville (1986) reviewed the current practice in designing durable structures. Although permeability was recognized as the key parameter, codes did not prescribe values for it. Instead, durability was prescribed by means of four 'Cs':

- constituents of the mix
- cover
- compaction
- curing.

In the ensuing discussion, a fifth 'C', cracking, was added. To make more direct provisions for the design life of a new structure Somerville argued that aggressive actions must be quantified and dealt with in a manner similar to strength, stiffness and serviceability. The owner should know how long the structure is required to remain in service; a performance plan is needed to predict whatever maintenance interventions will be required throughout its design life. This applies equally to evaluation of repair or protection options throughout the service life of an existing structure.

As Dawe (1990) pointed out, the main significance of a long prescribed life is to ensure that a structure is inspected and maintained in service, provided this is economic, for as long as it is required.

Leeming (1993) described the application of life cycle costing to bridges.

7.3 Selecting repair options

7.3.1 Typical options

Pritchard and Chubb (1987) reviewed the assessment of chloride-contaminated highway structures. An overall plan or strategy would be based on:

(*a*) data collection
(*b*) assessment
(*c*) investigation
(*d*) remedial action.

They presented a flow chart applicable to major structures when the investigation can be justified to avoid or delay expensive repairs or to minimize interference with traffic.

In the case of smaller structures the investigation would be short-circuited to move rapidly to repairs or remedial works.

Five options identified for evaluation in the maintenance of the Midland Links viaducts (see 2.3.6) were:

(i) do nothing
(ii) surface coating
(iii) cathodic protection
(iv) repair
(v) replace.

Wallbank (1989) preferred 'do minimum' followed by 'prevent leakage' in view of the importance of the latter.

Mays and Boam (6.1.7) subdivided repair into the repair of visible defects and major repairs.

7.3.2 Degradation mechanisms

Manning (1990) neatly summarized the common degradation mechanisms as shown in Figure 7.3 and mathematically as follows:

carbonation: $d = kt^{1/2}$

corrosion: $d = t_0 + kt^a$

where $a = 0.5$ for diffusion control or 1.0 for reaction control

sulphate attack: $d = t_0 + kt^b$

alkali-silica reaction: $d = t_0 + kt^b$ where $b > 1.0$

frost attack: $d = t_0 + kt$

where d is the deterioration,
t_0 is the initiation time,
t is time,
k is constant for a particular mechanism and concrete.

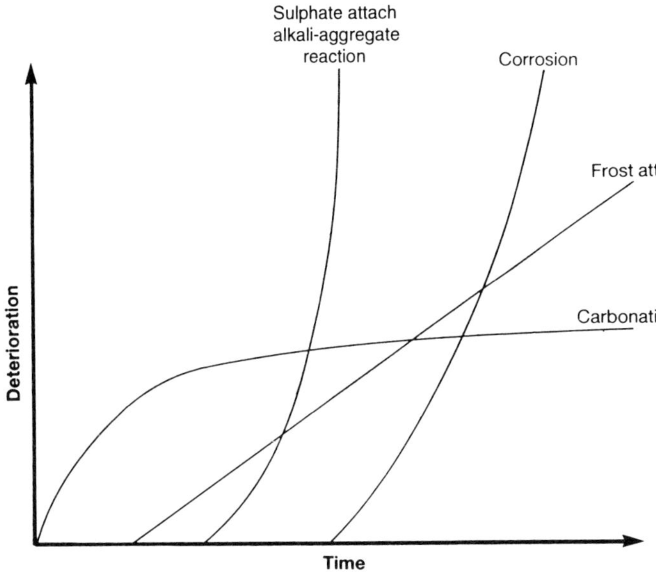

Fig. 7.3. Simplified degradation mechanisms (Manning, IABSE, 1990)

7.3.3 Rate of corrosion

US Federal Highway Administration (1983) issued information about the corrosion-inhibiting performance of epoxy-coated reinforcement and calcium nitrite admixture in accelerated tests. Results were compared with control specimens having black steel in concrete without admixture or salt. The slabs were cast in two pours with the bottom reinforcement in salt-free concrete and the top reinforcement in salty concrete. With only the top steel epoxy-coated, it took twelve times as long to consume iron equal to the control. If both mats were epoxy-coated, it took 46 times longer. The calcium nitrite admixture gave protection close to that of the epoxy-coated top mat.

When evaluating the needs of an existing structure, two questions arise:

(*a*) What is its present condition?
(*b*) What is its rate of deterioration?

Detection and monitoring methods were reviewed in Chapter 3.

Manning (1990) found visual inspection, half-cell potential mapping and sampling and testing methods time-consuming and expensive. In North America, impulse radar and infra-red thermography showed most promise for the rapid detection of delamination although neither detects general or localized pitting corrosion directly. To measure the rate of corrosion he favoured a method based on resistance polarization (also called linear polarization). For small potentials, normally 10 or 20 mV, the relationship between potential and current is linear with a slope known as the polarization resistance, R_p. The corrosion current is given by

$$i_{corr} = B/R_p$$

where B is a constant, depending on the anodic and cathodic Tafel slopes, with a value of 40 for most practical purposes. Experience had led to the following interpretations

i_{corr}	$< 0.20\ \text{mA/ft}^2$:	no corrosion damage expected
$0.20 < i_{corr}$	$< 1.0\ \text{mA/ft}^2$:	corrosion possible in 10–15 years
$1.0 < i_{corr}$	$< 10\ \text{mA/ft}^2$:	corrosion expected in 2–10 years
$i_{corr} >$	$10\ \text{mA/ft}^2$:	corrosion expected in less than 2 years.

This was useful information but structural inferences, with assumptions of general corrosion at a uniform rate, were only of limited value.

Wood (1990) had found that general corrosion could result in the loss of up to 1 mm per year from reinforcement section or up to 2 or 3 mm per year in pitting corrosion. He considered theoretical data and electrical monitoring to be erratic but he found that the selective exposure of steel in structures using high-pressure water jetting was beginning to provide an empirical basis for estimating the rate of corrosion.

Wood (1990) suggested that there are two reliable ways of determining permeability, K, and coefficient of diffusion, D, for concrete.

(i) Back calculation from service performance.
(ii) Short-term testing of concrete specimens.

The bulk diffusion test reported by Wood *et al.* (1989) accelerates diffusion ingress rates by a factor of 20. Wood had found that values of D for concrete at

25°C could range from 0.5 to over 500×10^{-12} m^2/s and values of K from 0.025 to over 250×10^{-12} m/s.

7.3.4 Criteria for concrete removal

Thompson (1989) found that repairing only visibly deteriorated concrete in a chloride-contaminated bridge deck (see 2.3.4) is not satisfactory. Within two months new anodic areas had appeared and, after three years, the half-cell potentials more negative than −350 mV (CSE) had increased from 25% to more than 40%.

Experience (2.2.4 and 2.3.4) shows that shallow patch repairs are seldom likely to be satisfactory. Where the reinforcement is corroded it is necessary to remove the concrete to behind the steel and to clean it thoroughly (4.3.2) or replace it if pitting is extensive or has reduced the cross-section significantly.

The Department of Transport (1990) requires concrete to be removed where the total chloride ion concentration at the level of the reinforcement exceeds 0.3% by weight of cement and half-cell potentials are more negative than −350 mV (CSE).

Wallbank (1989) proposed a chloride threshold of 0.2 or 0.15% by weight of cement in conjunction with half-cell potentials more negative than −350 mV (CSE).

Criteria for removing chloride-contaminated concrete have to be decided on the basis of an acceptably low risk of renewed corrosion.

7.3.5 Performance plans

In order to evaluate whole life costs of different repair or protection options, performance plans are drawn up, estimating all the future expenditure commitments of each option. Until adequate data banks are available, the estimates require careful judgement but the discipline of the method is clear. Options are compared on the basis of present value, applying discount factors appropriate to the owner's financial policy. Long-term discounting compresses the differences. White (1990) gave as an example the total estimated cost of maintaining a notional motorway bridge over 120 years. Today's Cost was £2,213,000 which, amortized at 3.81%, gave a Present Value of £929,047.

7.4 Specifications

7.4.1 Testing

BS6089 (1981). Assessment of concrete strength in existing structures.
BS1881 Testing concrete:

Part 114 (1983) Density of hardened concrete
Part 120 (1983) Cores
Part 121 (1983) Static modulus of elasticity in compression
Part 122 (1983) Water absorption
Part 124 (1988) Analysis of hardened concrete

Part 201 (1986) Guide to the use of NDT methods for hardened concrete
Part 202 (1986) Surface hardness
Part 203 (1986) Ultrasonic pulse velocity
Part 204 (1988) Electromagnetic covermeters
Part 205 (1986) Radiography
Part 206 (1986) Strain measurements
Part 207 (1992) Near to surface strength tests
Part 208 (in preparation) ISAT (revision of Part 5)
Part 209 (1990) Dynamic modulus of elasticity.

7.4.2 Patch repairs

Concrete Society (1991b). Technical Report 38.
Department of Transport (1986b). BD 27/86.

7.4.3 Flowable concrete

Department of Transport (1986b). BD 27/86.

7.4.4 Sprayed concrete

Concrete Society (1979) CS.021 and (1981) CS.022.
American Concrete Institute (1983). Revision of ACI 506–66.
American Society for Testing and Materials (1985). ASTM C820.
Japan Society of Civil Engineers (1984a) Guideline and (1984b) Specification.
Sprayed Concrete Association (1986). Codes of practice.
Department of Transport (1986b). BD 27/86.

7.4.5 Impregnation of concrete

Department of Transport (1990b). Criteria and material for the impregnation of concrete highway structures.
Departmental Standard BD 43/90.

7.4.6 Cathodic protection

Concrete Society (1989). Technical Report No. 36.
Concrete Society (1991a). Technical Report No. 37.

7.4.7 Alkali-silica reaction

Concrete Society (1987). Technical Report No. 30.

7.4.8 Structural adhesives

Mays and Hutchinson (1988). Engineering property requirements for structural adhesives. Compliance Spectrum for two-part cold cure adhesive for bonding steel to concrete. Mechanical properties of hardened adhesive and test methods.

BS5350 (1976). Methods of test for adhesives. Parts A–H.
BS6319 (1983). Testing of resin compositions for use in construction. Parts 1–10.

7.5 References

AMERICAN CONCRETE INSTITUTE (1983). *Recommended practice for shotcreting.* Revision of ACI Standard 506–66.

AMERICAN SOCIETY FOR TESTING AND MATERIALS (1985). *Standard specification for steel fibres for use in fibre reinforced concrete.* ASTM C820.

ANDRADE C., ALONSO C., GONZALEZ J. A. and RODRIGUEZ J. (1989). Remaining service life of corroding structures. *Symposium on the Durability of Structures.* Lisbon. International Association for Bridge and Structural Engineering, 359–364.

ARYA C. and NEWMAN J. B. (1990). Problem of predicting risk of corrosion of steel in chloride-contaminated concrete. Paper 9592. *Proc. Instn Civ. Engrs*, Part 1, **88**, Oct., 875–888.

BEEBY A. W. (1978). Cracking and corrosion. CIRIA/UEG. *Concrete in the Oceans Technical Report No. 1.* Department of Energy. Cement and Concrete Association, Slough.

BEEBY A. W. (1983). Cracking, cover and corrosion of reinforcement. *Concrete International*, **5**, No. 2, Feb., 35–40.

BEEBY A. W. (1990). G. Somerville (ed.). The contribution of research to the development of design for durability. British Group of IABSE Colloquium, Cambridge, July. *The design life of structures.* Blackie (an imprint of Chapman and Hall), Glasgow and London, 1992.

BLIGHT G. E. (1991). A study of four waterproofing systems for concrete. *Mag. Concr. Res.*, **43**, No. 156, 197–203.

BORGES J. FERRY (1989). Some basic concepts in building and their relationship to durability. *Symposium on the Durability of Structures.* Lisbon. International Association for Bridge and Structural Engineering.

BOYD W. K., LANKARD D. R. and SLATER J. E. (1979). *Influence of bridge deck repairs on corrosion of reinforcing steel.* Transportation Research Board, Washington DC, USA.

BROWN B. J. (1990). R. Dhir and J. W. Green (eds.) Silane treatment of concrete bridges. Protection of concrete. *Proc. Int. Conf. on Protection of Concrete* University of Dundee, 11–13 Sept., 1059–1080.

BROWN J. H. (1986). The performance of concrete in practice. *A field study of highway bridges.* TRRL Contractor Report 43.

BUCKLAND P. G. (1990). Canada's advanced national standard on bridge evaluation. *Proc. 1st Int. Conf. on Bridge Management.* University of Surrey, Guildford, 28–30 Mar., 575–584.

BURLEY E., RIGDEN S. and ABU-TAIR A. (1989). The performance of different repair techniques and materials under static loading. *Proc. Int. Conf. on Structural Faults and Repair.* Kensington, London, 27–29 June.

CABRERA J. G., CUSENS A. R. and LYNSDALE C. J. (1989). Porosity and permeability as indicators of concrete performance. *Symp. on the Durability of Structures.* Lisbon. International Association for Bridge and Structural Engineering.

CAIRNS J. (1990). An analysis of the behaviour of reinforced concrete beams following deterioration and repair. *Proc. 1st Int. Conf. on Bridge Management.* University of Surrey, Guildford, 28–30 Mar., 643–653.

CALDER A. J. J. (1986). *Repair of cracked reinforced concrete: assessment of injection methods.* TRRL Research Report 81.

CALDER A. J. J. and THOMPSON D. M. (1988). *Repair of cracked reinforced concrete: assessment of corrosion protection.* TRRL Research Report 150.

CALDER A. J. J. (1990). Corrosion protection afforded by various crack injection techniques. *Proc. 3rd Int. Conf. on Corrosion of Reinforcement in Concrete Construction.* Wishaw,

Warwickshire, 21–24 May. Published for the Society of Chemical Industry by Elsevier Applied Science, 433–445.

CIRIA (1977). *Rationalisation of safety and serviceability factors in structural codes.* Construction Industry Research and Information Association. Report 63.

CIRIA (1987). *Protection of reinforced concrete by surface treatments.* Construction Industry Research and Information Association, London. Technical Note 130.

CLEAR K. C. (1974). *Evaluation of Portland cement concrete for permanent bridge deck repair.* Federal Highway Administration. PB-232 604. National Technical Information Service. US Department of Commerce.

CLIMACO J. C. T. S. and REGAN P. E. (1989). Evaluation of bond strength between old and new concrete. *Proc. Int. Conf. on Structural Faults and Repair.* Kensington, London, 27–29 June.

CONCRETE SOCIETY (1979). *Specification for sprayed concrete.* CS.021, 4.

CONCRETE SOCIETY (1981). *Guidance notes on the measurement of sprayed concrete.* CS.022, 2.

CONCRETE SOCIETY (1982). *Non structural cracks in concrete: Report of Working Party.* Technical Report No. 22.

CONCRETE SOCIETY (1984). *Repair of concrete damaged by reinforcement corrosion: Report of Working Party.* Technical Report No. 26.

CONCRETE SOCIETY (1987). *Alkali-silica reaction. Minimising the risk of damage to concrete. Guidance notes and model specification clauses.* Technical Report No. 30.

CONCRETE SOCIETY (1989). *Cathodic protection of reinforced concrete: Report of a joint Concrete Society and Corrosion Engineering Association Working Party.* Technical Report No. 36.

CONCRETE SOCIETY (1990). *Assessment and repair of fire-damaged concrete structures.* Technical Report No. 33.

CONCRETE SOCIETY (1991a). *Model specification for cathodic protection of reinforced concrete.* Based on ICE Conditions of Contract. Technical Report No. 37.

CONCRETE SOCIETY (1991b). *Patch repair of reinforced concrete subject to reinforcement corrosion.* Model specification and method of measurement. Technical Report No. 38.

DARBY J. J. (1992). Maintenance—a County viewpoint. *Durability of highway structures Symp.* Institution of Highways and Transportation, Taunton, 12 Nov.

DAVIS A. G. and HERTLEIN B. H. E. (1990). A. S. Nowak (ed.). Assessment of bridge deck repairs by a non-destructive technique. *Bridge evaluation, repair and rehabilitation.* Nato ASI Series E: Applied Sciences, Vol. 187, ISBN 0 7923 0999 5.

DAWE P. H. (1990). G. Somerville (ed.). Management strategy and maintenance. British Group of IABSE Colloquium, Cambridge, July. *The design life of structures.* Blackie (an imprint of Chapman and Hall). Glasgow and London, 1992, 179–184.

DEPARTMENT OF TRANSPORT (1986a). *The investigation and repair of concrete highway structures.* Departmental Advice Note BA 23/86 (superseded by BA 35/90).

DEPARTMENT OF TRANSPORT (1986b). *Model specification for replacement concrete for the sides and soffits of beams and crossheads.* Departmental Standard BD 27/86.

DEPARTMENT OF TRANSPORT (1990). *Inspection and repair of concrete highway structures.* Departmental Advice Note BA 35/90.

DEPARTMENT OF TRANSPORT (1990b). *Criteria and material for the impregnation of concrete highway structures.* Departmental Standard BD 43/90.

EMBERSON N. K. and MAYS G. C. (1990a). R. K. Dhir and J. W. Green (eds). Design of patch repairs: measurements of physical and mechanical properties of repair systems for satisfactory structural performance. *Proc. Int. Conf. on Protection of Concrete.* University of Dundee, 11–13 Sept.

FEDERAL HIGHWAY ADMINISTRATION (1983). *Time to corrosion of reinforcing steel in concrete.* Report No. FHWA/RD-83/012.

HARSH S. and DARWIN D. (1984). SM Report 9. *Effects of traffic-induced vibrations on bridge deck repairs.* University of Kansas.

HARWOOD P. C. (1989). Surface coatings specification criteria. *Proc. 3rd Int. Conf. on Deterioration and Repair of Reinforced Concrete in the Arabian Gulf.* Bahrain Society of Engineers, Vol. 1, 635–646.

HERGENRÖDER M., SCHALL G. and RACKWITZ R. (1989). Reliability and risk function for deteriorating structures. *Symp. on the Durability of Structures.* Lisbon. International Association for Bridge and Structural Engineering.

HOPE B. B. and IP A. K. C. (1987). Chloride corrosion threshold in concrete. *ACI Materials J.* Title no. 84–M32, July–Aug.

IABSE (1989). *Symp. on the Durability of Structures.* Lisbon. International Association for Bridge and Structural Engineering.

IRWIN R. J. and CHAMBERLIN W. P. (1981). *Performance of bridge decks with 75 mm design cover.* Research Report NY/RR/-81/93. Federal Highway Administration. Washington DC, USA.

JACKSON P. A. and COPE R. J. (1990). Strength assessment for concrete bridges. *Proc. 1st Int. Conf. on Bridge Management.* University of Surrey, Guildford, 28–30 Mar., 429–438.

JAPAN SOCIETY OF CIVIL ENGINEERS (1984a). Guideline for construction of steel fibre reinforced concrete. *Concrete Library of JSCE,* No. 3, June, 31–39.

JAPAN SOCIETY OF CIVIL ENGINEERS (1984b). Specification of steel fibres for concrete. *Concrete Library of JSCE,* No. 3, June, 41–44.

KEER J. G., CHADWICK J. R. and THOMPSON D. M. (1990). Protection of reinforcement by concrete repair materials against chloride induced corrosion. *Proc. 3rd Int. Conf. on Corrosion of Reinforcement in Concrete Construction.* Wishaw, Warwickshire, 21–24 May. Published for the Society of Chemical Industry by Elsevier Applied Science.

KOJIMA T., MIYAGAWA T., NAKANO K., YAMAGUCHI Y. and KOBAYASHI K. (1992). Effect of coating to inhibit AAR in concrete structures. *9th Int. Conf. on AAR in Concrete.* Queen Elizabeth II International Conference Centre, London.

KUNISHIMA M. and OKAMURA H. (1989). Durability design for concrete structures. *Symp. on the Durability of Structures.* Lisbon. International Association for Bridge and Structural Engineering.

LAMBERT P. and PAGE C. L. (1986). The monitoring of reinforcement corrosion. *Proc. of Research Seminar: The Behaviour of Concrete Structures.* Cement and Concrete Association, 288–291.

LEEMING M. B. (1993). The application of life cycle costing to bridges. *Proc. 2nd Int. Conf. on Bridge Management.* University of Surrey, Guildford, 18–21 Apr.

LEFAS I. D. and KOTSOVOS M. D. (1989). Behaviour of repaired RC structural walls. *Proc. Int. Conf. on Structural Faults and Repair.* Kensington, London, 27–29 June.

MCKENZIE M. (1991). *Cathodic protection of reinforced concrete: effectiveness on initially uncorroded steel.* TRRL Research Report 328.

MACMILLAN G. L. and TREADAWAY K. W. J. (1990). An exposure trial of concrete durability in Arabian Gulf conditions. *Proc. 3rd Int. Conf. on Corrosion of Reinforcement in Concrete Construction.* Wishaw, Warwickshire, 21–24 May. Published for the Society of Chemical Industry by Elsevier Applied Science, 115.

MANNING D. G. *et al.* (1985). Debate: Crack width, cover and corrosion. *Concrete International,* **82**, No. 2, May, 20–35.

MANNING D. G. (1990). G. Somerville (ed.). Design life of concrete highway structures—the North American scene. British Group of IABSE Colloquium, Cambridge, July. *The design life of structures.* Blackie (an imprint of Chapman and Hall), Glasgow and London, 1992, 144–153.

MAYS G. C. (1983). *Fatigue strength of epoxy resin adhesive joints at elevated temperatures.* TRRL Contractor Report.

MAYS G. C. and HUTCHINSON A. R. (1988). Engineering property requirements for structural adhesives. *Proc. Instn Civ. Engrs,* Part 2, **85**, 485–501.

MIYAGAWA T. (1991). Durability design and repair of concrete structures: chloride corrosion of reinforcing steel and alkali-silica reaction. *Mag. Concr. Res.*, **43**, No. 156, Sept., 155–170.

MIYAGAWA T., HISADA M., INOUE S. and FUJII M. (1991). Effect of concrete surface treatment on expansion due to alkali-silica reaction. *Concrete Library of JSCE,* No. 18, Dec., 237–261.

MIYAGAWA T., INOUE S., FUJII M. and UEHATA N. (1992). Effect of concrete surface treatment on expansion due to alkali-silica reaction. *9th Int. Conf. on AAR in Concrete.* Queen Elizabeth II International Conference Centre, London.

OECD (1989). *Durability of concrete bridges.* Road Transport Research Group Report. Organisation for Economic Co-operation and Development, 2, rue André Pascal, 75775 Paris Cedex 16.

OECD (1992). *Bridge management.* Scientific Expert Group Report. Organisation for Economic Co-operation and Development, 2, rue André Pascal, 75775 Paris Cedex 16.

PEARSON-KIRK D. (1992). Impregnation of concrete. *Durability of Highway Structures Symp.* Institution of Highways and Transportation, Taunton, 12 Nov.

PIARC TECHNICAL COMMITTEE ON ROAD BRIDGES (1991). *Cathodic protection of bridge decks.* Drafted by Paolo Baldo. Permanent International Association of Road Congresses.

POINEAU D. (1989). Durability assessment from design phase to execution. *Symp. on the Durability of Structures.* Lisbon. International Association for Bridge and Structural Engineering.

PRITCHARD B. P. and CHUBB M. S. (1987). Concrete bridge integrity assessment. *Structural assessment—the use of full and large scale testing.* Butterworth, 9–19.

REEL R. S. and MURUGANANDAN C. (1990). Cost-effective strategies in bridge management. *Proc. 1st Int. Conf. on Bridge Management.* University of Surrey, Guildford, 28–30 Mar., 715–724.

REEL R. S. and GULIS E. (1993). Cathodic protection of bridge decks and piers in Ontario. *Proc. 2nd Int. Conf. on Bridge Management.* University of Surrey, Guildford, 18–21 Apr.

RIYADH J. A., MURJAN A. A., AMSO N. N. and ABBAS R. M. (1989). Performance of repaired RC beams subjected to static and cyclic loadings. *Proc. Int. Conf. on Structural Faults and Repair.* Kensington, London, 27–29 June.

ROBINS P. (1990). Fibre shotcrete repair applications in the UK. *British-Swedish Seminar on Concrete Repairs.* Chalmers University of Technology, Goteborg.

SALAH EL DIN O. A. and SALAMA A. E. (1989). Effect of fire temperature on the strength of concrete with different types of admixtures. *Proc. Int. Conf. on Structural Faults and Repair.* Kensington, London, 27–29 June.

SATOH Y., OHAMA Y., DEMURA K. and TACHIBANA K. (1989). Improvements in barrier properties of concrete by silane impregnation. *Proc. 3rd Int. Conf. on Deterioration and Repair of Reinforced Concrete in the Arabian Gulf.* Bahrain Society of Engineers. October, 689–699.

SCHIESSL P. (1988). *Corrosion of steel in concrete.* Report of Technical Committee 60–CSC RILEM, ISBN 0 412 32100 9.

SHARMA A. K. and ANDREWS G. (1989). Testing of repaired reinforced concrete beams. *Proc. Int. Conf. on Structural Faults and Repair.* Kensington, London, 27–29 June.

SIEMES T. and VROUWENVELDER T. (1985). Durability—a probabilistic approach. *International Association for Bridge and Structural Engineering Proceedings*, P-85/85, 65–76.

SINGH G. (1989). Life expectancy studies of reinforced concrete using a microcomputer. *Symp. on the Durability of Structures*. Lisbon. International Association for Bridge and Structural Engineering.

SOMERVILLE G. (1986). The design life of concrete structures. *The Structural Engineer*, 64A(2); Discussion 64A(9).

SPRAYED CONCRETE ASSOCIATION (1986). *Codes of practice for dry and wet process sprayed concrete*. October, London.

STANDING COMMITTEE ON STRUCTURAL SAFETY (1989). 8th Report. Construction Industry Research and Information Association.

STRATFULL R. F. (1974). *Experimental cathodic protection of a bridge deck*. Transportation Research Record No. 500, 1–15.

STRATFULL R. F. (1983). Criteria for the cathodic protection of bridge decks. Paper presented at meeting: *Corrosion of reinforcement in concrete construction*. London. 13–15 June. Society of Chemical Industry, 287–331.

THOMAS M. D. A. and MATTHEWS J. D. (1991). *Durability studies of PFA concrete structures*. BRE Information Paper IP 11/91.

THOMPSON D. M. (1989). Effectiveness of repair of reinforcement corrosion in a motorway viaduct. *Proc. Int. Conf. on Deterioration and Repair of Reinforced Concrete in the Arabian Gulf*. Bahrain Society of Engineers, 157–171.

THOMPSON D. M. (1990). Concrete repair in highway bridges. *Highways*, **58**, July, No. 1963, 19–21.

THOMPSON D. M. and LEEMING M. B. (1992). *Surface treatments for concrete highway bridges*. TRRL Research Report No. 345.

TRANSPORTATION RESEARCH BOARD (1981). *Concrete sealers for protection of bridge structures*. National Cooperative Highway Research Program Report 244.

TRITTHART J. (1990). *Chloride-binding in cement and the Cl/OH ratio of the pore solution*. Statens Vaeg och Trafikinstitut, Linkoeping. VTI Report No. 352a, 129–142.

VASSIE P. (1987). *The chloride concentration and resistivity of eight reinforced concrete bridge decks after 50 years' service*. TRRL Research Report 93.

VASSIE P. R. (1991). *The half-cell potential method of locating corroding reinforcement in concrete structures*. TRRL Application Guide 9.

WALLBANK E. J. (1989). *Performance of concrete in bridges—a study of 200 trunk road bridges*. HMSO, London. ISBN 011 550877 5.

WHITE K. H. (1990). G. Somerville (ed.). A performance approach to design. British Group of IABSE Colloquium, Cambridge, July. *The design life of structures*. Blackie (an imprint of Chapman and Hall), Glasgow and London, 1992, 30–39.

WOOD J. G. M., KING E. S. and LEEK D. S. (1989). Defining the properties of concrete repair materials for effective structural application. *Proc. Int. Conf. on Structural Faults and Repair*. Kensington, London, 27–29 June.

WOOD J. G. M., JOHNSON R. A. and ELLINAS C. (1990). Reliability analysis applied to deteriorating bridge structures. *Proc. 1st Int. Conf. on Bridge Management*. University of Surrey, Guildford, 28–30 Mar., 501–513.

WOOD J. G. M. (1990). G. Somerville (ed.). Predicting future decay in concrete structures. British Group of IABSE Colloquium, Cambridge, July. *The design life of structures*. Blackie (an imprint of Chapman and Hall), Glasgow and London, 1992, 185–194.

8

Conclusions

8.1 Durability and maintenance

8.1.1 Recognizing the problems

In the 1970s, growing awareness in many countries of the need to safeguard their stocks of bridges led to efforts towards:

(*a*) improving durability of concrete in highway structures,
(*b*) rehabilitation and protection of concrete bridges.

The OECD published a series of reports, and Bridge Management Systems (BMS) were developed in OECD member countries.

Research was concentrated upon the causes and mechanisms of deterioration and upon diagnostic and repair techniques.

Concrete bridges are required to maintain their serviceability over long periods of time, hopefully 100 years or more. In the USA (6.1.4), a bridge designed for durability is expected to become functionally obsolete before it becomes structurally deficient. The average age for replacement is 70 years and, during this lifetime, it would have been rehabilitated once.

8.1.2 Defects and their causes

Causes of deterioration or damage are:

- carbonation (1.2.1)
- chloride (1.2.2)
- sulphate (1.2.3)
- freeze-thaw (1.3.1)
- alkali-silica reaction (1.3.2)
- high alumina cement (1.3.3)
- impact (1.3.4)
- fire (1.3.5).

The most common defects (2.2, 2.3, 2.4 and 6.1) were:

- cracking, predominantly caused by restraint of early thermal or drying shrinkage,

- spalling due to reinforcement corrosion, attributed to lack of cover, chloride contamination, water leakage or the synergistic effect of carbonation and chloride (7.1.1, Figure 7.1).

The causes could often be traced back to:

- faulty design—details, drainage arrangements
- construction—insufficient depth or quality of cover
- lack of maintenance—leaking joints, blocked drains.

The case histories show that reinforcement corrosion, which is attributable mainly to chloride, either from sea water or road salt, can be serious in structures only 20 years old or even sooner in exceptionally aggressive environments such as that in the Arabian Gulf.

On the other hand, carbonation is unlikely to reach the steel, where the specified concrete cover is provided, for at least 40 years and possibly 80 years or more. This lends support to a conclusion of Brown (1986), that carbonation is unlikely to be a serious problem in modern bridges (see 6.1.7). He also noted the variability of concrete in the field. Durability problems tend to be associated with localized defects rather than the average quality of the concrete.

Deterioration of reinforced concrete bridge decks is usually due to ineffective waterproofing or leaking road joints or a combination of these with other causes. Substructures suffer from deicing salt leaking from above or being splashed up by traffic. Wetting and drying increases the concentration of chlorides and, in some cases, sulphates.

8.1.3 Looking for the answers

Investigations into the various deterioration mechanisms were undertaken at a number of research centres. By understanding the processes better, especially with regard to chloride attack, well informed decisions can be made when designing structures or managing existing structures. This applies especially to the evaluation of repair options (see 8.5.5) for which estimates have to be made of the useful lives of alternative proposals.

The interest of bridge maintenance engineers concentrated on the development of repair strategies to achieve the best value for money.

8.2 Detection and monitoring

8.2.1 Visual inspection

The first signs of deterioration are normally picked up in General or Principal Inspections. The defects in reinforced concrete are likely to be cracking, scaling, spalling or staining. The cause of cracking must be diagnosed. A great many cracks are caused in the early days after construction by the tensile stresses that result from the various restraints to shrinkage. Controlled structural cracking is inherent in reinforced concrete. The significance of cracks in relation to durability has been the subject of debate but the important distinction affecting repair is whether a crack is dormant or active. Defects are commonly due to more than one cause.

8.2.2 Concrete testing

The following tests are concerned with the concrete:

- covermeter (3.2.1)
- rebound hammer (3.2.2)
- ultrasonics (3.2.3)
- impact-echo (3.2.4)
- permeability (3.2.5)
- carbonation (3.2.6)
- thermography (3.2.7)
- impulse radar (3.2.8)
- slot cutting (3.2.9)
- coring (3.2.9)
- pull out tests (3.2.10)
- pull off tests (3.2.10).

Recent developments by the testing equipment makers are mostly concerned with automation, data logging and processing and monitoring systems

8.2.3 Reinforcement corrosion

Half-cell potential mapping (3.3.1) is carried out almost routinely by some organizations and was originally interpreted as shown in Table 8.1.

Table 8.1. Interpretation of half- cell potential

Half-cell potential: mV	Probability of corrosion: %
More negative than −350	95
Values from −200 to −350	50
Less negative than −200	5

However, the half-cell potential measurements are affected by environmental conditions, chloride concentrations etc. and could be misleading if not supported by other tests to define the areas to be reinstated to give a lasting repair. Chloride concentration may be the overriding criterion and complete replacement of a deck may be more economic in the long term than patching.

The order of the risk and rate of corrosion may be interpreted from resistivity measurement (3.3.2) as shown in Table 8.2.

Table 8.2. Interpretation of resistivity

Resistivity: Ωcm	Likelihood	Rate of corrosion
Greater than 12 000	Unlikely	Low
5000 to 12 000	Probable	Intermediate
Less than 5000	Almost certain	High

Chloride concentration (3.3.3) around the reinforcement may be the most important single indication of a serious condition but it is not yet possible to give criteria for its use alone. Various methods have been tried to measure the actual rate of corrosion but none is reliable or suitable for site testing.

It is more important to predict the initiation of corrosion which can be done most reliably on the basis of past service performance.

8.2.4 Post-tensioning tendons

The exploratory hole drilling technique is the most direct and practical method of inspecting post-tensioning ducts.

Radiography may be useful to give a picture of a fairly small area, showing the position and size of bars and tendons and an indication of where to drill, but it will not show corrosion or broken wires; the definition may be poor and it is expensive.

Defects in tendons could be very localized and anchorages are the most vulnerable and the most difficult locations to inspect. The following procedure is recommended.

- Assess the factors influencing corrosion and the sensitivity of the structure to loss of prestress.
- Inspect vulnerable areas, including drilling and contamination sampling.
- If prestress may have been lost, assess remaining prestress by coring or slotting of concrete (3.2.9, 3.5) or by drilling of reinforcement (3.5).
- If future durability is in doubt, install a monitoring system such as an optical fibre sensor technique.

In-situ stress can be determined by blind hole drilling (3.5) in concrete or in steel. Drilling the small hole in a wire is semi-destructive but the reduction in cross-section is only 8% for a 5 mm diameter wire or 4% for a 7 mm diameter wire. This degree of loss of strength in one or two wires of a 7-wire strand or a 12-wire tendon is not critical.

8.3 Repair techniques

8.3.1 Evolution of repair techniques (4.1)

The most pressing problem in North America (6.1.4) was the deterioration of unwaterproofed bridge decks and efforts were concentrated on the development of criteria for removing salt-contaminated concrete, various overlay materials and techniques.

In countries where bridge decks were waterproofed, the damage was less obvious, but when corrosion of the reinforcement did occur it was sometimes of the pitting variety which tended to concentrate research efforts onto the development of effective methods of detecting this hidden problem.

Repair techniques for bridges grew from traditional building practice using Portland cement mortar. Epoxy resins offered good strength, adhesion and impermeability but had thermal expansion and elastic moduli different from concrete. Toxicity of some hardeners was a health hazard that had to be guarded against. Polyester resins were given as an alternative which was intermediate in cost but had other disadvantages. Rubber or styrene-butadiene latices or polyvinyl acetate were used to modify the properties of cement mortar. In recent years, the advantages of condensed silica fume and pulverized fuel ash in making concrete more durable are being recognized in the field of concrete repairs.

Patch repair methods had a reasonable chance of success where corrosion was due to carbonation in the absence of chlorides. Where chlorides were present, such repairs would buy some time but not long-term durability.

However, carbonation is not a serious problem in modern bridges. If it causes spalling, this is generally superficial and the repairs are usually without structural significance. Chloride contamination, on the other hand, requires the removal of comparatively large areas of concrete.

8.3.2 Treatment of cracking (4.2)

Before deciding whether or how to repair cracking, its cause must first be determined. If monitoring shows that a crack is widening or lengthening, the structural cause of this has to be found and remedied. The FIP (1986), has given a classification of cracks in relation to their primary causes.

Injection of stabilized cracks may:

(*a*) restore structural integrity across the crack and/or
(*b*) block access of water and chloride ions to the steel.

The crack should be as clean as possible before filling, which is usually done by pressure injection but sometimes by ponding and gravity. The FIP (1991) distinguishes low pressure (up to about 2 N/mm^2) and high pressure (up to 30 N/mm^2) injection.

Epoxies were widely used in the USA (6.1.4) to inject cracks (also to coat reinforcement and prime substrates). UK research (6.2.2 and Table 7.1) found that injecting cracks with epoxy restored most of the structural integrity. Polyester and methylmethacrylate were not so effective.

8.3.3 Concrete removal (4.3.1)

The technique of hydrodemolition

- is good environmentally
- saves time
- is reasonable in cost
- causes no cracking of concrete
- leaves reinforcement undamaged and cleaned
- removes concrete selectively.

The FIP (1991) gives information on:

- hydraulic water jetting (10–40 N/mm^2)
- high-pressure water jetting (40–120 N/mm^2)
- hydro-jetting (140–240 N/mm^2)

with comprehensive advice on other mechanical methods and other aspects of preparing concrete surfaces.

8.3.4 Patch repairs

Mays and Wilkinson (1987) reported on structural performance of patch repairs (4.1.5) using:

- polyester resin mortar
- epoxy resin mortar with bonding agent
- epoxy resin mortar without bonding agent
- ordinary Portland cement mortar
- cementitious mortar modified with styrene-butadiene rubber polymer latex and SBR bonding agent.

Generally, the ultimate strength of the repaired test beams exceeded that of unrepaired controls. Deflections however were increased, reflecting the lower elastic modulus of the repair materials compared with the original concrete. The bond at the repair/substrate interface is of prime importance.

The successful performance of patch repairs depends very much on careful and understanding workmanship in the preparation as well as the application of the repair material. Materials with a mismatch in elastic properties may succeed by virtue of good adhesion and electrical insulating properties.

Surveys in New Zealand (6.1.2) showed that patch repairs met with varying success; failure was often due to insufficient care in cutting out and cleaning beforehand.

In North America (6.1.4), patch repairs were usually sealed as protection against corrosion and to improve appearance.

8.3.5 Sprayed concrete (4.3.5)

Reference should be made to the comprehensive Code of Practice published by the Sprayed Concrete Association.

Shotcrete was used quite widely in the USA (6.1.4) where some States required prequalification of operators.

8.3.6 Replacement concrete (4.3.4)

Blended cement concrete is recommended for bridge decks and high flow concrete for sides and soffits of beams.

In North America (6.1.4) concrete removal and replacement was practised by most bridge agencies, the majority using Portland cement concrete but some also used epoxy or latex-modified concrete.

8.3.7 External reinforcement (4.4)

Exceptionally, supplementary reinforcement may be added to a concrete member in the form of mild steel plates bonded by a suitable adhesive to the concrete surface. The Standing Committee on Structural Safety welcomed research undertaken at TRRL and other centres into design, application and the long-term durability of the bonding. SCOSS consistently advised that the technique should only be used with care and attention to detail.

8.3.8 Supplementary prestress (4.5)

The OECD (1983) and the FIP (1991) described two ways of applying supplementary prestressing:

- vertical or inclined tendons to augment shear resistance,
- longitudinal tendons to augment bending resistance.

The FIP also included strengthening by the relative displacement of supports to induce a self-equilibrated stress condition.

External unbonded post-tensioning tendons could be relatively easily inspected and replaced if necessary (2.5.1, 4, 10, 18, 19). On the other hand, post-tensioning tendons in ducts were more difficult to inspect (2.5.3, 5, 7, 8, 14, 15, 21) and the anchorages virtually impossible. In view of concerns about the durability of post-tensioned bridges, the Concrete Bridge Development Group (CBDG) is currently collating information on the pros and cons of external or grouted internal tendons. In the case histories (2.5) where tendons had corroded or fractured, the various solutions were:

- supplementary post-tensioning
- external bonded reinforcement
- demolition and reconstruction.

8.3.9 Structural and other aspects (4.7)

When planning repairs, structural aspects must be fully taken into account to safeguard the public, the site staff and the structure itself. If concrete removal will cause loss of bond or anchorage, temporary support may be required.

An essential part of any rehabilitation is dealing with the cause of deterioration, such as replacing the waterproofing, replacing or resealing joints and improving the drainage of surface water. Some simply supported viaduct spans (2.3.6) were linked at alternate piers, thereby halving the number of expansion joints.

In North America (6.1.4) overslabbing converted some short span simply supported viaducts into continuous structures with fewer joints through which salt laden water could drain.

Structures affected by alkali-silica reaction were treated according to the severity of the damage. In the early stages, it could be suppressed by surface treatment or encapsulation. If structural integrity was threatened, the structure was propped pending partial or, rarely, complete reconstruction.

Redecking was often a better option than patch repairs and had the advantage of providing an opportunity to upgrade the width or load carrying capacity.

Transverse load distribution can be improved by overslabbing.

Repair may be undertaken solely to make good deterioration or accidental damage but is often associated with, or occasioned by a need for strengthening. Various other standards may be involved or dominant such as new seismic requirements, dynamic response, care of historic structures and aesthetic aspects.

Access to the structure was an important consideration both for carrying out repairs and in making provision for future inspection and maintenance.

Fire damage was often less serious than at first it appeared to be, but one large structure had to be reconstructed after a fire.

8.3.10 Keeping traffic moving

Interference with traffic always has to be minimized and this forms an important, often dominant part of economic appraisal. Repairs to superstructures and repairs or part replacement of substructures have been carried out without any interference to traffic.

The key to a successful operation often lies in the choice of temporary works required to support the structure safely and maintain traffic with the minimum of restrictions.

Savings in time and cost can sometimes be made by coordinating traffic arrangements for bridgework with those required for road maintenance works.

8.4 Protection

8.4.1 Corrosion countermeasures (5.1)

Preventive measures start with checking the effectiveness of the bridge deck waterproofing layer, the sealing of joints and the disposal of surface water drainage. It may be possible to improve the control of deicing practices.

8.4.2 Concrete surface treatment (5.2)

Pfeiffer and Scali (1981) reported tests on 21 materials available for sealing concrete surfaces. Of these materials, two epoxies, a methylmethacrylate, a moist-cured urethane and an alkyl-alkoxy silane were clearly more effective in reducing the chloride ion content compared with an uncoated concrete specimen.

Treatments can be classified into three main groups:

- coatings/renderings
- pore liners/pore blockers
- sealers.

Coatings, which are applied like paints, and renderings, which are applied by trowel, rely on the formation of a surface barrier to the ingress of liquids and gases.

Pore liners and pore blockers are applied as low viscosity liquids which solidify by solvent evaporation.

Bijen (1989) reported the characteristics of the following more important resins:

- □ epoxies

- ☐ polyurethanes
- ☐ tar epoxies and tar polyurethanes
- ☐ polymethylmethacrylates.

He also described the characteristics of five groups of compounds for the impregnation of concrete (5.2.4).

There has been considerable interest in the effectiveness of surface treatment to inhibit ASR (see 5.2.5), one of the most promising being silane.

8.4.3 Coating of reinforcement (5.3)

Fusion-bonded epoxy-coated reinforcement (FBECR) was used increasingly in the USA for more than 20 years. However, since a notable failure and because of concern about the high level of permitted damage, confidence in FBECR has been shaken and it is unlikely to find much application in bridge repairs.

Galvanized bars are considered to be more corrosion-resistant than uncoated bars in carbonated concrete but not where there is chloride contamination.

To reduce the incidence of new corrosion sites adjacent to repairs, Bijen (1989) proposed that reinforcement in repair areas should be insulated by drying and applying coats of

- ☐ solvent-free or low-solvent epoxy,
- ☐ epoxy with zinc extender or
- ☐ polymer-modified cement.

8.4.4 Cathodic protection

Cathodic protection can stop corrosion in salt-contaminated bridge decks regardless of the chloride content. Therefore only visible spalling and delaminated concrete have to be repaired. There is no need to remove large volumes of concrete because it is contaminated with chloride. Cathodic protection will prevent future corrosion including incipient corrosion at the boundaries of repaired areas.

The following systems are available:

- ☐ Deck anodes:
  - conductive asphalts
  - slotted systems
  - conductive mesh systems.
- ☐ Substructure anodes:
  - conductive mesh systems
  - conductive coatings
  - discrete anodes
  - sprayed metal coatings.

8.4.5 Desalination and realkalization (5.5)

Equipment similar to that used for cathodic protection is required to pass a current through the concrete using an external anode mesh attached to the concrete surface.

In desalination, chloride ions migrate to the surface anode and are removed. In realkalization, the impressed current forces alkali ions, particularly sodium, to the cathodic reinforcement where the reaction converts water and oxygen into hydroxyl ions and repassivates the steel. However, the process greatly increases the permeability of the concrete and all of the chloride is not removed. It is therefore advisable to follow up with silane impregnation.

8.5 Performance

8.5.1 Performance of bridges

Brown (1986) (6.1.7) found that the quality of concrete in a sample of modern British highway bridges was superior to that of older ones. However, many of the older bridges performed well, probably due to having no expansion joints and being on roads that are not heavily salted.

Wallbank (1989) (6.1.9) reported that three-quarters of all deck joints were leaking and 72% of bridges were found to have chloride concentrations exceeding 0.2% by weight of cement. He concluded that some 30% of English motorway and trunk road bridges were in poor condition on the basis of visual inspection. Rather more of the bridges showed spalling and this deterioration was attributed to the various causes in the following percentages:

chloride	12
freeze/thaw	11
impact	9
relative movement or other causes	6
	38

In the USA (6.1.4), 18% of highway bridges were considered obsolescent and another 24% defective.

In Belgium (6.1.8), about 50% of concrete bridges suffered from cracking or reinforcement corrosion.

About 10% of highway bridges in India (6.1.13) needed replacement and a further 10% were showing signs of distress.

In the former Yugoslavia, 19% of bridges (6.1.10) were judged to be in poor condition.

Swedish National Road Administration bridges (6.1.5) were reported to be in relatively good condition which was ascribed to design philosophy, regular inspections and the systematic bridge management systems which had been applied.

PFA concretes had greater strength, lower permeability, greater resistance to the ingress of chlorides and better suppression of alkali-silica reaction than OPC concrete.

8.5.2 Performance of repairs (7.1.3)

Broadly, the properties of materials with adequate compressive and tensile strength compared with those of typical bridge concrete as follows.

- Resin mortars had high flexural and tensile strengths but tended to have lower elastic moduli.
- Polymer-modified cementitious materials had coefficients of thermal expansion/contraction much the same as those of many unmodified concretes.
- Cementitious materials were more compatible with the substrate concrete.

As cracks provide preferential means of ingress of aggressive agents to the reinforcement, durability will be enhanced by sealing them effectively. In addition, injecting cracks with epoxy, polyester or methylmethacrylate resins could restore about half of the stiffness loss due to cracking.

Keer, Chadwick and Thompson (1990) presented a ranking of the protection of reinforcement by twelve repair materials compared with control specimens of good and moderate quality concrete.

8.5.3 Performance of surface treatments (7.1.4 and 8.4.2)

Silane impregnation was an effective aid to the drying out of concrete affected by alkali-silica reaction.

Thompson and Leeming (1992) reported water vapour, water and chloride ion transmission of 21 surface treatments and postulated general performance requirements.

8.5.4 Performance of cathodic protection (7.1.5)

In North America, cathodic protection of concrete bridge decks has gained general acceptance as a fully proven technique. Cathodic protection has also been applied to substructures there since the mid 1980s, but not widely.

In the UK, the major interest lies in its potential applications to substructures. It seems likely that quite high installation costs could be outweighed by considerable savings in the extent of physical repairs to chloride-contaminated reinforced concrete. Alternative anode systems are on trial.

8.5.5 Considering options (7.3.1)

Some or all of the following options are typical:

- do minimum
- prevent leakage
- surface treatment
- repair visible defects
- cathodic protection
- major repair
- partial or complete replacement.

Computer software has been developed (3.3.6) to evaluate various strategies and find the 'least cost' option.

Two questions arise:

- ☐ What is the present condition?
- ☐ What is the rate of deterioration?

To answer these, use can be made of well established models (7.3.2) supported by the best service data available.

To evaluate the whole life costs of different repair or protection options, performance plans (7.3.5) are drawn up, estimating all the future expenditure commitments of each option and expressing them at their present value.

8.5.6 Reliability (7.2.1)

There has been considerable interest in the adaptation of reliability methods to repairs. The full advantages of reliability methods will only be realized when enough service data have been accumulated.

Somerville (1993) is coordinating a BRITE project, involving two Spanish and three Swedish partners. The objective is illustrated in Figure 8.1.

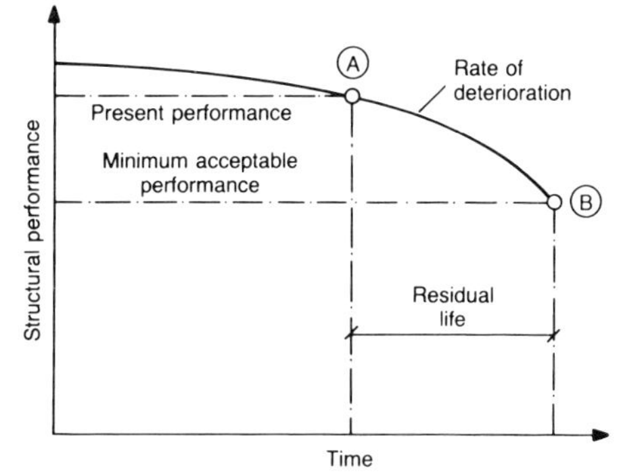

Fig. 8.1. Deterioration of structural performance (Somerville, BCA, 1993)

Starting from point A, the rate of deterioration needs to be predicted to establish point B. The mechanisms considered, singly and in combination, are:

- ☐ corrosion
- ☐ freeze-thaw and
- ☐ alkali-silica reaction.

The project is one-third complete and the partners wish to contact cooperative owners in order to test their assessment methods in a series of case studies.

8.6 References

BIJEN J. M. J. M. (1989). Maintenance and repair of concrete structures. *Heron*, **34**, No. 2, 49–63.

BROWN J. H. (1986). The performance of concrete in practice. *A field study of highway bridges.* TRRL Contractor Report 43.

FIP (1986). *Inspection and maintenance of reinforced and prestressed concrete structures.* Fédération Internationale de la Précontrainte. Thomas Telford, London.

FIP (1991). Repair and strengthening of concrete structures. *Guide to good practice.* Fédération Internationale de la Précontrainte. Thomas Telford, London.

KEER J. G., CHADWICK J. R. and THOMPSON D. M. (1990). Protection of reinforcement by concrete repair materials against chloride induced corrosion. *Proc. 3rd Int. Conf. on Corrosion of Reinforcement in Concrete Construction.* Wishaw, Warwickshire, 21–24 May. Published for the Society of Chemical Industry by Elsevier Applied Science.

MAYS G. and WILKINSON W. (1987). Polymer repairs to concrete: their influence on structural performance. *Katharine and Bryant Mather Int. Conf. on Concrete Durability.* Detroit, American Concrete Institute, SP-100.

OECD (1983). *Bridge rehabilitation and strengthening.* Road Transport Research Group Report. Organisation for Economic Co-operation and Development, 2, rue André Pascal, 75775 Paris Cedex 16.

PFEIFFER D.W. and SCALI M. J. (1981). *Concrete sealers for protection of bridge structures.* National Cooperative Highway Research Program Report 244. Transportation Research Board, Washington DC, USA.

SOMERVILLE G. (1993). *Residual service life of concrete structures.* BRITE project BREU-CT92-0591. British Cement Association. *Research Focus*, No. 13, Apr.

THOMPSON D. M. and LEEMING M. B. (1992). *Surface treatments for concrete highway bridges.* TRRL Research Report 345.

WALLBANK E. J. (1989). Performance of concrete in bridges — a study of 200 trunk road bridges. HMSO, London. ISBN 011 550877 5.

Appendix I. Historic concrete bridges

Built	Ref.	Bridge	Spans: m	Repaired
1870	2.2.1	Flixton Estate, River Waveney	16.5, iron framed	1993
1875	(3)	Seaton, River Axe	10, 16.2, 10, unreinforced	
1894	(2)	Gefion footbridge, Copenhagen	17°skew 19, reinforced concrete arch	First in 1966
1897	(3)	Glenfinnan railway viaduct, Scotland	300, overall in 21 spans, unreinforced	
1901	2.2.6	Zuoz, River Inn	38, reinforced concrete	Redecked 1969
1905	(1)	Tavanasa, River Rhine	52, reinforced concrete arch	
1909	2.2.16	Mizen Head, County Cork	52, reinforced concrete arch	1980
1912	2.2.18	Woodbridge, Guildford	21, reinforced concrete arch	1960 and 1988
1914	2.2.15	Sázava River near Kácov	3, reinforced concrete spans continuous	
1915	(3)	Newark, Nottinghamshire footbridge	27, reinforced concrete arch	
1926	2.6.12	Limmat Valley, Baden-Wettingen	72, reinforced concrete arch	1976
1926	2.2.2	Railway bridge, River Cam	13, river and 8 of 7 reinforced	1974
1928	2.2.3	Royal Tweed, Berwick	51, 75, 87 and 110, reinforced	1980
1946	(3)	Adam railway viaduct, Wigan	9.2, precast prestressed	
1947	(3)	Nunn's, Fishtoft, Lincolnshire.	22.5, in-situ post-tensioned	

AI References

(1) BILLINGTON D. P. (1979). *Robert Maillart's bridges — the art of engineering.* Princeton University Press, New Jersey, USA and Guildford, UK.

(2) BRØNDUM-NIELSEN T. (1966). *Professor Ostenfeld's Centenary. Bygningsstatiske Meddelelser.* 13 Oct. Published by Danish Society for Structural Science and Engineering.

(3) STANLEY C. C. (1979). *Highlights in the history of concrete.* Cement and Concrete Association (British Cement Association).

Appendix 2. Some early lightweight aggregate concrete bridges

Built	Ref.	Bridge	Construction
1966–67	(1)	Dyckerhoff Footbridge, Wiesbaden	LWAC 64.2 m central portion of 96.4 m span arch
1967–68	(1)	Hermannsweg Bridge, Hansalinie motorway	85 m span double cantilever with central shear-hinge
1967–68	(1)	Harmsen Viaduct, Rotterdam	LWAC 26.8 m central portion of 50.7 m long three-span structure
1967–68	(1)	Solleks River Bridge, NW Washington	Precast LWAC girders and long raking struts in three-span road bridge 70 m long overall
1969	(2)	Hanley Footbridge, Stoke-on-Trent	In-situ LWA/natural fines in ramps
1969	(2)	Redesdale Forest, Northumberland	LWAC pre-tensioned precast beams with LWAC in-situ infill
1970–71	(1)	Fühlinger See Bridge, near Cologne	LWAC 84 m central length of the post-tensioned three-span structure 240 m long overall constructed in situ in segments
1971–72	(1)	Industrial bridge, Wiesbaden	LWAC 78 m long precast longitudinal and transverse post-tensioned girders and in-situ deck slab
1971–72	(1)	Dukenburgse Bridge, near Nijmegen	LWAC 105.4 m central portion constructed in cantilever in situ in segments
1972	(2)	Southlands Bridge, Margaretting bypass	LWAC in-situ deck slab
1973	(2)	Saddlebow Bridge, King's Lynn bypass	LWAC in-situ deck slab
1973	(2)	Glasshouse Wood FB, Kenilworth bypass	LWAC in-situ portal
1976	(2)	Little Mill FB, East Peckham, Kent	Precast LWAC T-beam
1977–78	(2)	Friarton Bridge, Perth, Scotland	In-situ LWAC deck slab

A2 References

(1) Cembureau (1974). *Lightweight aggregate concrete technology and world applications*. The European Cement Association.

(2) Mays G. C. and Barnes R. A. (1991). The performance of lightweight aggregate concrete structures in service. *The Structural Engineer*, 15 Oct., **69**, No. 20, 351–361.

Appendix 3. Demolition

Sometimes investigations and evaluation result in the decision to demolish and reconstruct. More often, structures become functionally obsolete, notably by reason of widening projects. Demolition calls for an understanding and responsible approach and is supported by technological developments in drilling, thermic lancing, hydrodemolition (4.3.1) and explosives.

Work has to be carefully planned, never more so than in the case of post-tensioned concrete. Lindsell (1975) gave details of procedures adopted in the demolition of nine prestressed bridge beams. The FIP *Guides to good practice* deal with the subject generally (1982) and in the context of repair (1991). Guidance was issued by the Health and Safety Commission (1979), the British Standards Institution (1982) and the Health and Safety Executive (1984a and b, 1985 and 1988).

The PIARC (1987) Technical Committee on Bridges, reviewed methods of removing concrete for repairs (4.3.1).

New Civil Engineer (1991) reported that most of the demolition at Wardley Hall Viaduct (2.3.10) was done with jackhammers but an ultra-high pressure water jetting system operating at 248 N/mm^2 was also used.

New Civil Engineer (1992b) reviewed the use of explosives to remove concrete bridges in a motorway widening contract. The preparatory hole drilling can be done with little effect on traffic. Then, during a planned closure of the motorway for, typically, twelve hours, charges are placed, a mat of chippings laid on the road below the bridge, the charges fired and the debris removed after any secondary breaking that may be required.

New Civil Engineer (1992a) reported the removal of Ingst Bridge over the M4 motorway in Avon which had been demolished after 25 years. As in-situ demolition by explosives or piecemeal methods was not feasible, it was decided to lift the 79 m long 2045 t superstructure out in one piece and transport it 2 km along the motorway and thence to a specially widened layby for breaking up. It was lowered by four multi-wheeled transporter jacks and, after a 36-hour delay due to one of the steerable bogies becoming stuck, the move was made.

New Civil Engineer (1992b) compared this operation with a more successful one in which the A1 trunk road traffic was carried on a temporary bridge. This enabled the contractor to demolish the redundant underbridge, slice it into manageable pieces with diamond impregnated steel wire, remove the pieces and construct the new bridge with minimal interruption of traffic.

New Civil Engineer (1993) reported the use of diamond saws in the removal of poor quality concrete from a new bridge.

A3 References

BRITISH STANDARDS INSTITUTION (1982). *Code of Practice for demolition.* BSI, London.

FIP (1982). Demolition of reinforced and prestressed concrete structures. *Guide to good practice.*

FIP (1991). Repair and strengthening of concrete structures. *Guide to good practice.* Thomas Telford, London.

HEALTH AND SAFETY COMMISSION (1979). *Safety in demolition work.* Construction Industry Advisory Committee. ISBN 0118832425.

HEALTH AND SAFETY EXECUTIVE (1984a). *Health and safety in demolition work.* GS29/2, Part 2, Legislation. ISBN 0118835890.

HEALTH AND SAFETY EXECUTIVE (1984b). *Health and safety in demolition work.* GS29/3, Part 3, Techniques. ISBN 0118836099.

HEALTH AND SAFETY EXECUTIVE (1985). *Health and safety in demolition work.* GS29/4, Part 4, Health hazards. ISBN 0118836048.

HEALTH AND SAFETY EXECUTIVE (1988). *Health and safety in demolition work.* GS29/1, Part 1 (revised), Preparation and planning. ISBN 011854054.

LINDSELL P. (1975). Demolition of post-tensioned concrete. *Concrete,* Jan., **9**, No. 1, 22–25.

NEW CIVIL ENGINEER (1991). Wardley Hall Viaduct piers replaced. *New Civil Engineer,* 1 Aug. 1991.

NEW CIVIL ENGINEER (1992a). (M4 Ingst Bridge removal). No regrets over M4 saga. Clear the decks. *New Civil Engineer,* 5 Mar. 1992, 5, 22–23.

NEW CIVIL ENGINEER (1992b). Big bang theory. *New Civil Engineer,* 13 Aug. 1992, 29–30.

NEW CIVIL ENGINEER (1993). Amey cuts up new M4 bridge. *New Civil Engineer,* 29 Apr. 1993, 5.

PIARC TECHNICAL COMMITTEE ON BRIDGES (1987). Section 4. *Repair and rehabilitation of bridge decks.* Permanent International Association of Road Congresses. 27, rue Guenegaud—75006 Paris.

